Overview of world elasmobranch fisheries

FAO FISHERIES TECHNICAL PAPER

341

by
Ramón Bonfil
Instituto Nacional de la Pesca
Progreso, Yucatán, Mexico

Food and Agriculture Organization of the United Nations

Rome, 1994

The designations employed and the presentation of material in this publication do not imply the expression of any opinion whatsoever on the part of the Food and Agriculture Organization of the United Nations concerning the legal status of any country, territory, city or area or of its authorities, or concerning the delimitation of its frontiers or boundaries.

M-43
ISBN 92-5-103566-0

All rights reserved. No part of this publication may be reproduced, stored in a retrieval system, or transmitted in any form or by any means, electronic, mechanical, photocopying or otherwise, without the prior permission of the copyright owner. Applications for such permission, with a statement of the purpose and extent of the reproduction, should be addressed to the Director, Publications Division, Food and Agriculture Organization of the United Nations, Viale delle Terme di Caracalla, 00100 Rome, Italy.

© **FAO 1994**

PREPARATION OF DOCUMENT

Elasmobranchs are interesting both from a scientific perspective and because of the characteristics their biology poses for their management. They may be associated with other major fisheries and often fail to get the attention they deserve. This publication is a contribution to reducing the oversight that is so often the misfortune of this group. We hope that the overview and detailed regional descriptions will help both the worker at the regional level as well as those involved in overall syntheses.

Distribution:

FAO Fisheries Department
FAO Regional Fishery Officers
Marine Sciences (General)
Author

Overview of World Elasmobranch Fisheries

FAO
FISHERIES
TECHNICAL
PAPER
341

ERRATUM

p. iv

Bonfil, R.
Overview of World Elasmobranch Fisheries.
<u>FAO Fisheries Technical Paper</u>. No 341, Rome, FAO. 1994. 119p.

ABSTRACT

Major world fisheries for elasmobranchs are described in regard to their importance, recent trends, problems for assessment and management, conservation and the outlook for their sustainability. The analysis considers trends and outlooks in FAO's statistical areas, individual accounts of fisheries by the major elasmobranch fishing nations and the high seas fisheries that have significant by catches of elasmobranchs. It appears that elasmobranch catches could be increased significantly in Areas 87, 67, and 47, whereas Areas 21, 27, 37, 61 and 71 probably will not sustain much further expansion in elasmobranch yields. The world elasmobranch catch in 1991 was 704 000t; if present trends continue it could reach 755 000-827 000t by the year 2000. However, the total annual catch inclusive of discarded and unreported catches is estimated at around 1.35 million tonnes. According to reported catches from the last 15 years, sharks account for almost 60% of the world elasmobranch catch while skates and rays make almost 40%. Twenty six countries have major elasmobranch fisheries on the basis that their present or recent catches of sharks and rays exceed 10 000t/yr. Information on species, gears, patterns of exploitation, research and management of elasmobranchs is summarized for each of these countries. The general problems in appraising and managing elasmobranch fisheries and the need for conservation are discussed and possible solutions for some of these problems are proposed.

---

p. 98

The figure annotations ETP and WCP have been transposed.

ISBN 92-5-103566-0   ISSN 0429-9345

ROME, 1994

ACKNOWLEDGEMENTS

Access to otherwise unobtainable published and unpublished data and information on the particular elasmobranch fisheries of their countries/regions is gratefully acknowledged to the following persons: Mr. Leonardo Castillo, Instituto Nacional de la Pesca, Mexico City, Mexico; Dr. Che-Tsung Chen, National Taiwan Ocean University; Dr. Pauline Dayaratne, National Aquatic Resources Agency, Colombo, Sri Lanka; Mr. Shigeto Hase, North Pacific Anadromous Fish Commission, Vancouver, Canada; Dr. David Holts, National Marine Fisheries Service, La Jolla, USA; Dr. Rosangela Lessa, Universidade Federal Rural do Pernanbuco, Recife, Brazil; Mr. Julio Morón, Indo-Pacific Tuna Development and Management Programme, Colombo, Sri Lanka; Dr. Ramón Muñoz-Chápuli, University of Málaga, Spain; Dr. Sigmund Myklevoll, Institute of Marine Research, Bergen, Norway; Mr. Larry J. Paul, MAF Fisheries, Wellington, N.Z.; Ms. Chee Phaik Ean, Fisheries Research Institute at Penang, Malaysia; Dr. Andrew Richards, Mr. Paul Tauriki and Mr. Paul V. Nichols, Forum Fisheries Agency, Honiara, Solomon Islands; Mr. Pairoj Saikliang, Depart of Fisheries, Bangkok, Thailand; Dr. Carolus M. Vooren, Universidade do Rio Grande, Brazil; Ms. Pouchamarn Wongsawang, Southeast Asian Fisheries Development Center Samutprakarn, Thailand.

Dr. Timothy A. Lawson, South Pacific Commission, Noumea, New Caledonia, kindly contributed maps for some figures. Dr. Tony Pitcher gave useful ideas on early drafts of this document. Dr. David Die, FAO Rome, provided helpful comments to this document as well as valuable data. Dr. Kent Carpenter and Dr. Richard Grainger, FAO, Rome provided helpful technical assistance and gave access to the FAO fisheries statistical database.

Ms. Alida Bundy kindly gave useful comments to the final version of the text. My final thanks go to Mrs. Ratana Chuenpagdee for all her assistance in the collection of data, preparation of figures and tables and final editing and typing of the review.

CONTENTS

1. INTRODUCTION. 1
   1.1 Background to Elasmobranch Fisheries 1
   1.2 A Note on Taxonomy 2

2. WORLD ELASMOBRANCH FISHERIES. 4
   2.1 The Official Statistics 4
      2.1.1 Catches by FAO Major Fishing Areas 4
      2.1.2 Catches by countries 7
   2.2 Major Fisheries for Elasmobranchs 8
      2.2.1 America 12
         2.2.1.1 USA 12
         2.2.1.2 Mexico 20
         2.2.1.3 Peru 24
         2.2.1.4 Brazil 24
         2.2.1.5 Argentina 27
      2.2.2 Europe 28
         2.2.2.1 Norway 28
         2.2.2.2 Former USSR 31
         2.2.2.3 United Kingdom 32
         2.2.2.4 Ireland 34
         2.2.2.5 France 36
         2.2.2.6 Spain 38
         2.2.2.7 Italy 39
      2.2.3 Africa and Indian subcontinent 40
         2.2.3.1 Nigeria 40
         2.2.3.2 Pakistan 41
         2.2.3.3 India 43
         2.2.3.4 Sri Lanka 44
      2.2.4 Asia 46
         2.2.4.1 Japan 46
         2.2.4.2 South Korea 47
         2.2.4.3 People's Republic of China 49
         2.2.4.4 Taiwan (Prov. of China) 51
         2.2.4.5 Philippines 53
         2.2.4.6 Thailand 54
         2.2.4.7 Malaysia 56
         2.2.4.8 Indonesia 58
      2.2.5 Australian subcontinent 59
         2.2.5.1 Australia 59
         2.2.5.2 New Zealand 61
   2.3 By catches and Discards of Elasmobranchs at Sea 63
      2.3.1 Drift gillnet fisheries 64
         2.3.1.1. North Pacific Ocean 64
            Salmon fishery 64
            Flying squid fishery 67
            Large-mesh driftnet fishery 72

    2.3.1.2 South Pacific Ocean 72
    2.3.1.3 Indian Ocean 75
    2.3.1.4 Atlantic Ocean 75
    2.3.1.5 Overview of driftnet fisheries 78
  2.3.2 Longline fisheries 79
    2.3.2.1 Atlantic Ocean 79
    2.3.2.2 Indian Ocean 87
    2.3.2.3 Tropical and South Pacific 91
    2.3.2.4 North Pacific 94
    2.3.2.5 Overview of longline fisheries 95
  2.3.3 Purse seine fisheries 97
  2.3.4 Other miscellaneous fisheries 99
  2.3.5 Overview 100
    2.3.5.1 Species of elasmobranchs under pressure from high seas fisheries. 101

3. DISCUSSION 102
 3.1 Current Situation of Elasmobranch Fisheries 102
 3.2 Problems for the Assessment and Management of Elasmobranch Fisheries . . 103
  3.2.1 Biology and Fisheries Theory 103
  3.2.2 Multiplicity of Species and Gears 104
  3.2.3 Economics, Shark "finning" and Baseline information 104
 3.3 Conservation of Elasmobranchs 104

4. SUMMARY AND CONCLUSIONS 106

5. BIBLIOGRAPHY 107

1. INTRODUCTION

1.1 Background to Elasmobranch Fisheries

As a group, elasmobranchs present an array of problems for fisheries management and conservation. Their life-history characteristics make them a fragile resource, more susceptible to overfishing than most teleost fishes. Assumptions of traditional fisheries models do not always fit the biological traits of elasmobranchs, making their assessment and management difficult. The high mobility of many species, sometimes involving trans-boundary migrations, incorporates another level of complexity to their assessment and highlights the need for proper knowledge about stock delimitation and dynamics if adequate management to be implemented. Elasmobranch fisheries assessment is complicated further because of a general lack of baseline information about their fisheries throughout the world. Furthermore, the economics driving elasmobranch exploitation involve a paradox: sharks and rays have a relatively low economic value making them low priority resources when it comes to research or conservation, while the demand for some of their products, such as shark fins, is very high and stimulates increased exploitation. The demand for shark fins sometimes results in substantial waste when only the fins are kept and the rest of the fish is discarded.

Considering these circumstances, it is not surprising that there is a history of non-sustainability in the exploitation of elasmobranchs. In recent years however, there has been growing international concern over the conservation of some elasmobranch stocks and it seems that now, more than ever, there is a need for a more systematic approach to the problem of elasmobranch assessment and management.

Fisheries for elasmobranchs have not increased in the same way as because of other fisheries worldwide. The low market value of these fishes, and their relatively low abundance. Compagno (1990) indicates that in terms of the commercial catches and according to FAO statistics, cartilaginous fishes are a minor group which contributed an average of 0.8% of the total world fishery landings during 1947-1985, while bony fishes such as clupeoids, gadoids and scombroids, accounted for 24.6%, 13.9% and 6.5%, respectively. Furthermore, elasmobranch catches increased only threefold over this time whilst the other three groups showed fivefold to sixfold increases and total world catches increased fourfold. Recorded world chondrichthyan commercial catches totalled 704 000t in 1991 (present study) making 0.7% of the total world fisheries catches; even considering an unreported catch of 50% to those recorded, this is still only about 1% of the world fisheries catch. Despite these facts, elasmobranch resources are of prime importance in some regions of the world and have sustained very important fisheries in some countries. Also, they have been, and remain, a cheap source of protein for millions of humans from coastal communities dependent on subsistence fisheries.

Traditionally, elasmobranchs have not been a highly priced fishery product. Their economic value ranks low among marine commercial fisheries (e.g. in the Taiwanese gillnet fisheries of the Central Western Pacific, shark (trunks) prices attain only 20% and 60% of those of tunas and mackerels (both whole) respectively (Millington 1981)). Exceptions are: sport fisheries, which can be of considerable economic value; certain species for whom a gastronomic demand has recently developed in some parts of the world (e.g., mako and thresher sharks in USA), or those species which, unfortunately, are highly-sought only for their teeth and jaws, such as the great white shark. The only highly-prized elasmobranch product is shark fin for oriental soup, a commodity for which there has recently been a considerable increase in demand (Cook 1990). On the other hand, anthropocentric points of view have substantially biassed public opinion against some elasmobranchs labelling them either as malevolent or as trash fish and thus undesirable species. A further issue is that of shark attacks on humans and the damage that some sharks cause to fishing gear and catches. These problems are real, but now are probably insignificant compared to the threats that humans represent to some populations of elasmobranchs.

Sharks and rays have biological characteristics and an ecological role which suggests they could be particularly vulnerable to fishing pressure. All elasmobranchs are predators and most exist at the top of the food chain. Their abundance is therefore relatively small compared to groups situated in lower trophic levels. They are typically slow growing and long-lived and mature at a late age. This, together with their low fecundity, results in a low reproductive potential for most of the species. Recoveries of population numbers from severe depletions (caused either by natural phenomena or human action) should take many years for the majority of elasmobranchs. Additionally, the removal of top predators from marine ecosystems might trigger undesirable consequences for the environment and other fishery resources (van del Elst 1979).

The vulnerability of the group, together with the past history of collapses in elasmobranch fisheries (see Anderson (1990) for a review), are causes for concern. The continuing increase in their catches and the continuing increase in demand for shark fins may be endangering the sustainability of elasmobranch fisheries. However, adoption of any widespread conservation measure is likely to affect the fisheries of many countries for whom the resource is of considerable importance. These impacts are difficult to assess without good basic information about their fisheries on a global scale.

Given the relative low value of elasmobranchs it is not surprising that information on their fisheries, or even their basic biology, is scarce, patchy and scattered, especially when compared to the amount of literature on other fishery resources or even that focusing on the problem of incidental catches of marine mammals in fisheries. An example is the amount of scientific literature generated during the past 16 years. The results of a fisheries database (ASFA) query for papers published between 1978 and 1993 including the name of six different fishery resources in the article's title are shown in Figure 1.1. Sharks and rays rank last after the salmons, shrimps and prawns, clupeoids, tunas and even lobsters. Although these figures may be biassed in some cases (e.g. inclusion of numerous environmental studies for salmonids and aquaculture studies for shrimps and prawns), they still are a measure of the importance of each resource.

To properly assess the current situation of elasmobranch resources, address the various problems associated with their exploitation and contribute new ideas about their study and management, it is essential to increase our knowledge about the characteristics and diversity of their fisheries, the species exploited, the size of the catches, discards at sea and past or current management measures adopted for the fisheries. While recent workshops and symposia have expanded our knowledge, specially in relation to their biology, much of the existing information about their fisheries is not only dispersed but is also not usually published by those concerned with the studies or management. This review is a contribution towards providing this information by compiling in a single volume the most important information available about world elasmobranch fisheries and providing an analysis of the global situation.

1.2   A Note on Taxonomy

Elasmobranchs are part of the Chondrichthyes. The Class Chondrichthyes comprises a diverse group of fishes whose most obvious common feature is the possession of a cartilaginous skeleton as opposed to the bony skeleton of the Osteichthyes or bony fishes. The cartilaginous fishes form an ancient and successful group dating back to the Devonian era. The basic models remain largely unchanged since their last large speciation during the Cretaceous era. Despite

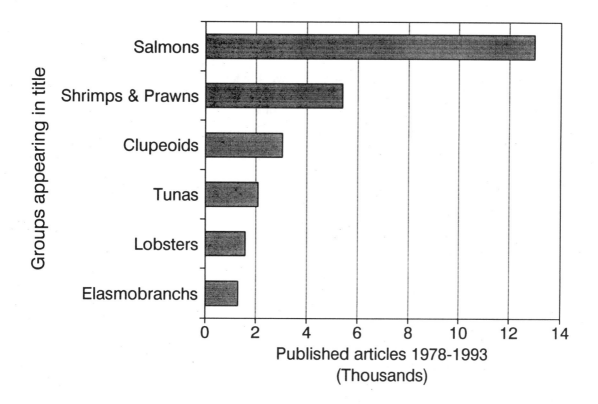

Figure 1.1. Number of articles published* during 1978-1993 for each of six groups of fishery resources, and appearing in the ASFA database. (* those with the name of the group in the article's title)

their ancient origin they possess some of the most acute and remarkable senses found in the animal kingdom which allows them to coexist successfully with the more modern teleost species.

The chondrichtyans are grouped into two main subclasses by many taxonomists: Holocephalii (Chimaeras or ratfishes and elephant fishes) with three families and approximately 37 species inhabiting deep cool waters; and the Elasmobranchii, which is a large, diverse group (including sharks and rays) with representatives in all types of environments, from fresh waters to the bottom of marine trenches and from polar regions to warm tropical waters. The great majority of the commercially important species of chondrichtyans are elasmobranchs. The latter are named for their plated gills which communicate to the exterior by 5-7 openings. The classification of elasmobranchs is a subject of continuous debate but they are generally divided into three groups, i.e., squalomorphs, galeomorphs and squatinomorphs, which include 30 families and approximately 368 species and a group known as the batoids composed of rays, skates, torpedoes and sawfishes, embracing 14-21 families and about 470 species (Compagno 1977, 1984; Springer and Gold 1989).

For this review, all the Chondrichthyes, (sharks, skates, rays and chimaeras) are often treated together as "elasmobranchs" or "sharks and rays". Although this is strictly inaccurate, it simplifies writing and reading by avoiding uncommon or lengthy terminology such as "chondrichthians" or "sharks, skates, rays and chimaeras" every time reference is made to the group.

2. CHARACTERIZATION OF ELASMOBRANCH FISHERIES

Organization of this work

The first section examines official statistics worldwide to describe the scale of global elasmobranch fishing. The section consists of an overview of the catch statistics by FAO major fishing areas including short-term projected catches and an overview of the trends in the most important fisheries for elasmobranchs in the world on a country basis. In the next two sections a more detailed analysis of elasmobranch fisheries is given. For this review, countries with reported elasmobranch catches of 10 000t/yr or more are called "major" elasmobranch-fishing countries.

The second and third sections deal with the major fisheries for elasmobranchs, the by catches and their discards at sea. Although it is difficult to distinguish between directed and incidental fisheries, especially when dealing with fishes that are seldom targeted and/or caught alone as is the case of sharks and rays, these two main divisions of elasmobranch commercial fisheries are used. Directed fisheries are taken as those that target elasmobranchs, together with coastal or small scale multispecies fisheries which catch elasmobranchs incidentally. Typically, the catches from these two sources are mixed together in the official statistics of most countries and it becomes necessary to treat them together. But, there is a group of large-scale long-range fisheries that mainly target high value species such as tunas which catch elasmobranchs incidentally and which mostly discard them for various reasons. These fisheries comprise an essentially different category in which the elasmobranch resource is not only wasted, but the actual numbers of elasmobranchs caught are also poorly known and usually do not form part of reported fisheries statistics. Most cases in this category are high seas large scale fisheries with driftnets and longlines, carried out by a few countries and targeting very specific resources such as tunas, billfishes, salmonids and squid. These fisheries are suspected of causing substantial mortalities of elasmobranchs, mainly sharks. This has raised concern over the conservation of these fishes, though it is secondary to concern over marine mammals, which are also frequently taken as by catch. Depending on the amount of information available in each case, the species, catches, gears, fishing units, localities, levels of exploitation and existing management or conservation measures, are summarized.

2.1 The Official Statistics

The data used in this analysis was taken from official fishery statistics of each country. The first source was the compilation of Compagno (1990) who analyzed FAO data for the period 1947-1985. FAO figures since 1970 have been updated using their Fisheries Yearbooks for 1988-1991 (FAO 1990-1993 and data provided directly from the FAO statistical database (David Die, FAO, pers. comm. August 2, 1993). Additional sources are: Fishery Statistical Bulletins for the South China Sea Area years 1976-1990 (SEAFDEC 1977-1993, appendix 1), the Fisheries Yearbook of Taiwan Area for 1970 and 1988-1990 and the Mexican Fishery Statistical Yearbooks 1976-1990 (Secretaria de Pesca 1979-1992, appendix 1). After the review of FAO data by Compagno (1990), the information is updated here and expanded, including, in particular, the catches of Taiwan (Prov. of China).

2.1.1 Catches by FAO Major Fishing Areas

Total world elasmobranch catches reported for the period 1947-1991 (Figure 2.1) grew to a record 704 000t in 1991. Roughly four periods with different trends can be identified. Poor growth in catches between 1947 and 1954, a sustained increase of production during 1955-1973 followed by a period of sluggish production for most of the 70's and finally renewed growth in catches during the last years 1984-1991.

Catches by major FAO Fishing Areas from 1967 to 1991 are summarized in Table 2.1. An attempt is made to rank these areas according to their catch. Because the sizes, coastline lengths and human populations of each area vary notably, a rough index of relative production was devised for comparison. This index is defined as the average total elasmobranch catch of each area divided by the square root of the surface of the area in km². A better index might have been the size of the continental shelf for each area but it was not possible to obtain these data. Arbitrarily, values of the index below 5 were considered indicative of low relative production, those between 5 and 10 intermediate and those of more than 10. as high. Additionally, the trend in catches during the last 10 years recorded for each area is expressed as the slope of a least squares linear regression.

In the Western Atlantic Ocean, all the areas have fairly high increasing trends, especially Area 21 (North West Atlantic) which has the most rapidly increasing trend in the world. All three areas show strong variations in their catches. Area 21 had the highest variability with recent years apparently recovering production from a dramatic drop suffered in the late 70's following high yields in the early 70's. Area 21 had a marginally higher index of relative production (IRP), but considering that a good part of this area includes arctic waters practically void for fishing, a much higher future IRP should not be expected from this area. In the Western Central Atlantic (Area 31) there was of a moderate increase in catch trend while the IRP indicated a low elasmobranch yield. This agrees with Stevenson (1982) who suggested that elasmobranch resources in this area could have been under-utilized. Perhaps there is still potential for expansion of catches, mainly for countries of the Caribbean region. For Area 41 (South Western Atlantic), elasmobranch catches also show a moderate increasing trend after variable catches in the 60's. Average catch of elasmobranchs in Area 41 is the highest in the Western Atlantic but this is also the largest area. Thus it has only an intermediate IRP. Small increases in catches might still be possible here in the future. Catches in Area 31 have been the lowest in the Western Atlantic while in the first half of the period and during the last two years, Area 21 had the highest yields.

For the Eastern Atlantic Ocean, Area 27 (the North Eastern Atlantic) had by far the largest catches in the Atlantic as well as the third largest and the second least variable catches in the world. According to the IRP this area has the highest production of elasmobranchs worldwide but further expansions in the catches should probably not be expected. In fact, the catch trend hardly increased as production has fallen since 1988, perhaps showing that the high levels of exploitation in this area are not sustainable. The Central Eastern Atlantic (Area 34) shows a medium variation in elasmobranch production. This area increased its catches during the early 1970s but the recent trend is of a slow decline. This is an area with an intermediate IRP, thus a good recovery in catches could be possible. For the Mediterranean Sea (Area 37), production was relatively variable during the period examined. The recent trend of declining catches is the steepest in the world. Because of the small size and the high density of human settlements of this Area, fishing is intense and the IRP for elasmobranchs is the second highest in the Atlantic Ocean. Very likely, elasmobranchs stocks here are close to full exploitation. In Area 47 (South Eastern Atlantic) catches have been fairly variable. It has the second smallest mean catch of elasmobranchs and the lowest IRP in the world, showing the most possibilities for increased exploitation of elasmobranchs in the future. For the four areas of the Eastern Atlantic, Area 27 dominated the catches producing more than the other three areas together.

Table 2.1. Elasmobranch catches by FAO Statistical Area 1967-1991. Mean catch, variation and index of Relative Production (IRP) are given for the last 25 years, and catch trends for the last 10 years.

F A O Major Fishing Areas	Area Million Km2	Mean Catch '000 t	Coefficient of Variation	I.R.P. Avg Catch/SqrtArea	Trend 82-91 '000 t/y
27 NE Atlantic Ocean	16.9	94.8	12%	23.07	0.26
61 NW Pacific Ocean	20.5	102.3	10%	22.60	-0.29
51 W Indian Ocean	30.2	97.6	19%	17.75	1.16
21 NW Atlantic Ocean	5.2	26.5	57%	11.61	5.48
37 Mediterranean & Black Seas	3.0	18.2	29%	10.50	-0.76
71 W Central Pacific Ocean	33.2	59.1	38%	10.26	5.00
41 SW Atlantic Ocean	17.6	34.2	30%	8.15	0.60
57 E Indian Ocean	29.8	42.9	32%	7.87	1.34
34 E Central Atlantic Ocean	14.0	28.6	29%	7.63	-0.65
87 SE Pacific Ocean	16.6	21.4	32%	5.24	-0.39
31 W Central Atlantic Ocean	14.7	17.4	47%	4.54	0.77
77 E Central Pacific Ocean	57.5	21.1	34%	2.79	0.08
81 SW Pacific Ocean	33.2	10.4	47%	1.81	0.55
67 NE Pacific Ocean	7.5	4.8	60%	1.74	0.20
47 SE Atlantic Ocean	18.6	6.6	42%	1.53	0.07

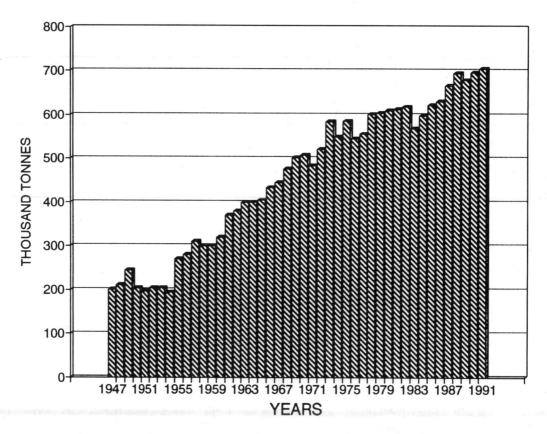

Figure 2.1 World reported catch of elasmobranch fishes 1947-1991 (Data from FAO and SEAFDEC, Fishery Yearbooks for Taiwan Area, and Secretaría de Pesca).

There are only two FAO areas in the Indian Ocean. The Western Indian Ocean (Area 51) has the second highest average yield in the world. This area has shown reasonably low variability in catches but a decreasing trend in recent production. Catches increased steadily up to the early 1970s but fell dramatically during 1983. Judging from the recent increasing trend in production, the situation seems to be recovering but catches have not yet reached previous levels. The IRP of Area 51 is the third highest in the world. Most of the catches in this area are taken in the northern region by Pakistan, India and Sri Lanka. Stocks in the northern region might be close to over-exploitation but given the large extension of this area and the low catches from its southern portion it might present some possibilities for increasing elasmobranch exploitation especially those of oceanic species. Area 57 (Eastern Indian Ocean) shows more variable catches with a growing trend. It has an intermediate IRP and higher yields are expected here. In the Indian Ocean, Area 51 produces, on average, more than double the catches of Area 57.

In the Western Pacific Ocean, Area 61 (North Eastern Pacific) had a decreasing trend for recent catches and the lowest variability of elasmobranch catches in the world. This area had the highest average yields in the world and the IRP was accordingly very high, marginally second to the North Eastern Atlantic. Therefore, stocks in this area might not provide any substantial increases in the future and may even be over-exploited. Area 71 (Central Western Pacific) showed the second highest trend of increase in catches reaching in the last few years five times those of the mid-1960s. The IRP in this area is relatively high and may indicate that yields could probably not be expanded much more. In the South Eastern Pacific (Area 81) catches have varied substantially with a low positive trend of recent catches. Average catches and therefore the IRP are very low. One possible reason for this is the relatively small extension of coastline inside this area together with correspondingly few human settlements. The potential of this area for significantly increased catches will depend mainly on the abilities of the stocks of oceanic and deep water elasmobranch species to sustain fisheries. Of the three areas of the Western Pacific, Area 61 is the most important for elasmobranchs having produced on average almost twice the catch of Area 71 and about ten times that of Area 81.

For the three areas of the Eastern Pacific, Area 67 (North Easter Pacific) has the smallest average catch and the highest variation in the world. The IRP is the second smallest in the world and the trend of recent catches is moderately positive. Larger catches might be obtained here in the future. Area 77 (Central Eastern Pacific) has variable catches with a low increasing trend and low IRP. Area 77 is the largest in the world but its low population density might account for the low IRP. The potential for increasing catches here is probably good especially in Central American countries and in the vast oceanic waters. The South Eastern Pacific (Area 87) is the only area of the East Pacific with a negative trend in catches and has an intermediate IRP. Further increases in the catches should be possible. Of the whole Eastern Pacific, Areas 77 and 87 have almost the same average catch during this period amounting to about four times those of Area 67.

Assuming that recent trends will continue without major changes in each of the FAO fishing areas, reported catches of elasmobranchs in the world can be expected to reach between 755 000t and 827 000t by the year 2000. These forecasts are based on "jackknife" linear regression analyses of elasmobranch catches since 1967 in each FAO major fishing area using a step of 5 years.

2.1.2 Catches by countries

Data by countries for the period 1947-1991 indicate that 26 countries presently harvest, or have recently harvested, more than 10 000t/yr of elasmobranch fishes. These countries are often referred to as "major elasmobranch-fishing countries". The elasmobranch catches of the People's Republic of China, although not available, also surpass 10 000t/yr and China is included as one of these 26 countries.

Catch statistics for the 25 major elasmobranch fishing countries for which data are available are shown in Table 2.2. Japan has traditionally been the overall major fisher of elasmobranchs in the world with average catches of 65 000t/yr. Indonesia, India, Taiwan (Prov. of China) and Pakistan follow with catches between 33 000t/yr and 43 000t/yr. France, the UK, the former USSR and Norway, recorded between 21 000t/yr and 27 000t/yr. Mexico, Brazil, South Korea, Nigeria, Philippines, Sri-Lanka and Peru caught between 11 000t/yr and 18 000t/yr. A large group formed by Spain, USA, Malaysia, Argentina, Thailand, Australia, Italy, New Zealand and Ireland followed with average catches between 4 000 and 10 000t/yr.

Even though there is great variability in the development of individual elasmobranch fisheries some patterns can be identified. About one third of the major elasmobranch fishing countries show recent levelling in their catches, probably signalling full exploitation of their resources. Seven countries show falling trends while nine others have a definite rise in catches (Figure 2.2).

Elasmobranch production is specially high in Indonesia where catches have soared since the early 1970s with no sign of a slow-down. Taiwan (Prov. of China), the USA, Spain and India are other countries with increasing landings of sharks and rays. Japan, historically the leader in elasmobranch fishing, has a clear trend of decreasing catches. Norway showed a clear increasing trend until the early 60's but catches have since sharply decreased. The same is true for the former USSR catches which grew from the early 60's to the mid-1970s but have since substantially decreased with no recovery. Catches in the UK have a very slight decreasing trend. Pakistan had a strong increasing trend in catch until the late 1970s, but dramatically dropped in the early 80's to be followed by a slow but sustained comeback. The range of causes for these decreasing trends is not easy to find in all cases but possible explanations for some cases follow.

The reported statistics indicate that during the last 15 years sharks have been slightly more important in catches than other elasmobranchs. The average reported catch of sharks and batoids is 285 433t/yr and 180 196t/yr respectively with an additional 190 159t reported as "various elasmobranchs." After reallocating catches wrongly reported as "various elasmobranchs" to either sharks or rays with the help of ancillary information and splitting the remaining 94 139t/yr of "various elasmobranchs" in equal parts, a total of 393 741t/yr (about 59.5% of total elasmobranchs) can be attributed to sharks whereas 262 046t/yr are skates and rays (about 39.5%). Less than 1% are quimaeras and elephant fishes.

2.2 Major Fisheries for Elasmobranchs

Two main sources provided the information in this section. First, literature on the subject was consulted for each case as extensively as possible. Much information probably remains in the form of unpublished reports from different governmental offices. Second, in an attempt to fill in some of the many gaps of information, a questionnaire was sent to officers or scientists in all major elasmobranch fishing countries. However, the success of this approach was poor. The extent of published work on elasmobranchs in each country and the level of response to the questionnaire is reflected in the quantity of information that is presented under each country's account.

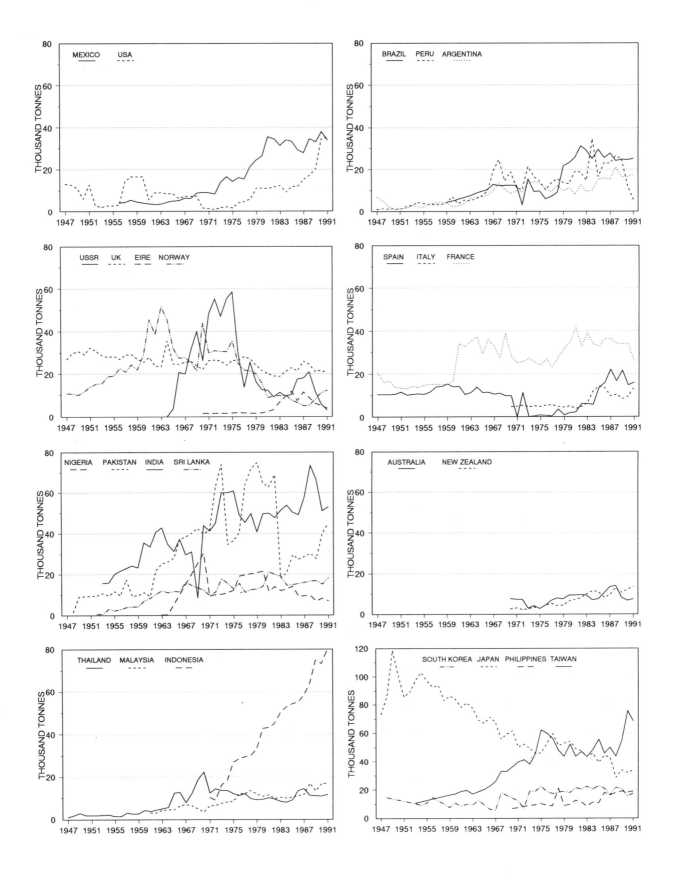

Figure 2.2. Historical catches of elasmobranchs for the 25 major elasmobranch fishing countries arranged by geographical area

Table 2.2. Commercial elasmobranch fisheries, reported world catches in thousand tonnes (data from Compagno, 1990 and FAO, unless otherwise indicated) (T.W.F. = total world fisheries, T.W.CUPL = total world cupleoid fisheries, T.ELAS = total world elasmobranch fisheries, EL/FISH = T.ELAS as % of TWF, CUPL/FISH = T.ELAS as % of T.W.CUPL).

YEAR	T.W.F.	T.W.CUPL	T.ELAS.	EL/FISH %	CUPL/FISH %	USA	MEX (p)	BRA	PERU	ARG	USSR	UK	EIRE	NORW
1947	20000	3481	201	1.0	17.4	13.1			1	6.9		27.1		10.8
1948	19600	3486	211	1.1	17.8	12.8			1.4	5.1		29.8		10.7
1949	20100	3724	245	1.2	18.5	11.2			1.2	2.4		30.7		10
1950	21100	4081	204	1.0	19.3	6.1			1.3	1		29.2		12
1951	23600	4392	197	0.8	18.6	12.8			1.1	1.2		32.6		14
1952	25200	5440	203	0.8	21.6	3.1			2.5	1.7		30.8		15.3
1953	25900	5500	204	0.8	21.2	2			3	2.9		28.8		15.5
1954	27600	5760	194	0.7	20.9	2.8			4.5	2.4		27.8		18.8
1955	28900	6410	270	0.9	22.2	2.8				2.2		28.6		19.1
1956	30500	7020	280	0.9	23.0	3.3	4.1		3.3	3.8		27.1		22.8
1957	31500	7230	310	1.0	23.0	14.3	4.5		3.5	4.1		29.1		20.9
1958	32800	7450	300	0.9	22.7	16.6	5.6		3.4	4.6		29.2		24.4
1959	36400	9060	300	0.8	24.9	16.6	4.6	4.6	4.2	4		27.2		22
1960	39500	10290	320	0.8	26.1	16.6		5	7.2	2.4		25.7		29
1961	43000	12620	370	0.9	29.3	5.7	3.6	5.9	3.8	2.9		27.8		45.6
1962	46400	14730	380	0.8	31.7	9	3.4		5.4	3.9		23.6		38.7
1963	47600	14930	400	0.8	31.4	9	3.5	7.6	5.1	6.2		23.5		51.6
1964	52000	18730	400	0.8	36.0	8.6	4.4	8.9	6.1	6.9	0.1	35.7		45.7
1965	52400	17442	405	0.8	33.3	8.6	5.1		7.6	7.2	3.7	24.7		32.2
1966	57300	19426	433	0.8	33.9	6.3	5.3	10.6	9.9	7.7	20.8	24.5		27.6
1967	60400	20308	444	0.7	33.6	7.3	6.5	13	19.6	10.1	20.1	25.6		27.7
1968	63900	21117	476	0.7	33.0	7.3	6.3	12.5	24.7	13.7	31.9	25.9		25.3
1969	62700	18786	502	0.8	30.0	7.3	8.9		14.7	10.8	40.1	23.8		21.5
1970	70388	22209	508	0.7	31.6	1.7	9.1	12.6	19	8.7	26.3	22.3	1.7	44.1
1971	70747	20241	482	0.7	28.6	1.5	9	12.6	11.3	10	48.3	26.3	1.7	29.8
1972	66121	14288	519	0.8	21.6	1	8.4	3.2	10.5	9.6	55.3	26.6	1.5	31.1
1973	62824	12073	583	0.9	19.2	1.8	14.1	15.6	21.5	13.4	47.1	26	1.5	30.5
1974	66597	14631	549	0.8	22.0	2.2	16.6	9.5	16.8	14.3	55.3	24.1	1.7	30.6
1975	66487	14373	586	0.9	21.6	1.7	14.3	9.9	14.6	13.8	58.5	26.5	1.8	35.9
1976	69930	15371	544	0.8	22.0	4.1	16.1	6.1	10.5	10.6	29.4	26.6	1.9	24.8
1977	69226	13043	556	0.8	18.8	4.7	15.6	7.3	13.8	9.6	13.7	28.1	1.8	21.9
1978	70596	14493	600	0.9	20.5	5.9	21.5	9.3	15.6	12.5	25.7	27.2	1.5	21.5
1979	71331	15790	603	0.8	22.1	11.1	24.6	21.9	13.8	10.0	16.2	24.2	1.7	20.0
1980	72141	16070	609	0.8	22.3	11.2	26.6	23.3	13.3	11.3	12.6	21.6	1.8	15.6
1981	74884	16920	612	0.8	22.6	11.0	35.7	25.8	19.1	8.3	12.5	20.3	2.5	8.9
1982	76810	17867	617	0.8	23.3	11.7	34.6	31.3	18.8	12.8	9.2	18.9	3.2	9.6
1983	77591	17455	568	0.7	22.5	12.4	31.4	29.1	14.9	9.5	11.2	18.8	6.8	9.8
1984	83989	19607	598	0.7	23.3	9.3	34.1	25.2	34.4	10.2	9.5	21.2	9.4	10.1
1985	86454	21101	623	0.7	24.4	11.9	33.3	29.6	16.8	15.3	10.2	23.0	11.8	7.8
1986	92822	23955	630	0.7	25.8	12.1	29.4	25.7	23.3	16.1	17.5	21.5	7.3	6.5
1987	94379	22375	666	0.7	23.7	15.2	27.9	27.8	23.1	15.3	18.1	25.9	11.4	5.1
1988	99016	24388	694	0.7	24.6	17.2	34.6	24.3	26.6	21.1	20.9	24.6	8.9	5.2
1989	100208	24800	679	0.7	24.7	20.4	33.1	24.9	25.0	16.5	12.0	21.2	6.2	8.0
1990	97434	22183	695	0.7	22.8	34.6	38.1	24.7	12.6	16.7	6.0	21.7	5.0	11.1
1991	96926	21407	704	0.7	22.1	35.5	34.0	25.2	5.7	17.6	3.1	20.4	4.0	12.3
MEAN	57896	14357	455	0.8	24.4	9.8	17.4	16.4	11.7	8.8	22.7	25.7	4.3	21.4
%variation	43	46	37	14	20	76	72	55	72	59	75	14	80	56
% of worldwide elasmobranch catch, 1987-1991						3.57	4.88	3.69	2.71	2.54	1.75	3.31	1.03	1.21
% importance of elasmobranchs in country, 1987-1991						0.42	2.36	3.00	0.29	3.19	0.11	2.63	3.03	0.44

(p) data from Secretaria de Pesca (Appendix 1)
(s) data from SEAFDEC (Appendix 1)
(s/f) data from SEAFDEC and FAO (Appendix 1)
(t/f) data from Fishery Yearbooks for Taiwan Area and FAO

Table 2.2.   Continued

SPAIN	ITALY	FRA	NIG	PKST	INDIA	SRILK	THAI (s)	MALAY (s)	INDONE (s/f)	S KOR	JAPAN	PHILIPP (s)	TAIWAN (t/f)	AUST	N ZEL
10.4		20.5					1				73.2				
10.4		16		1.5			2			14.6	86.1				
10.6		16.7		9.1			3				118.5				
10.8		13.7					2				100.7				
11.6		13.5					2				85.7				
10.1		13.1		9.8		0.6	2				89.1				
10.8		14.4		10.8	15.9	0.7	2.2			10.5	97.4		10.7		
10.9		13.7		9.8	16	3.1	2.3			9.2	102.9				
10.8		14.9		11.7	20.4	2.5	1.6			10.8	97.2				
11.7		15.2		9.7	21.9	3	1.6			14.8	92.6				
14.1		15.2		17.6	23.1	3.9	3.1			12.2	93.8				
14.2		15.2		9.5	24.3	4.3	2.7			10.2	82.9				
15.4		15.1		9.8	23.5	4.3	2.8			7.6	86		16.5		
14		16.7		11.3	35.6	7.1	4.3			10.9	83.9		17.1		
14.3		34.3		9.4	33.6	8.5	4	3.2		8.7	78.3		18.9		
10.6		33.1		22	40.8	10.3	4.5	3.2		9.9	81.5		19.7		
11.4		35.5	0.3	25.2	43	12.1	5.1	4.4		9.4	77.4		17.1		
13.8		37.4	0.3	26.2	34.9	11.2	5.8	4.7		12.6	69		18.8		
11.4		29.5		28.2	31.4	11.8	12.4	4.6			66.9		20.2		
11.5		36.3		37.2	37.4	11.6	12.8	6.4		6.3	71.1		22.9		
10.8		33.1		38.4	29.6	16.3	8	7		5.6	67.5		26.0		
11.1		27.4		40.3	31.2	14.7	12.3	6.5		18	56		33.1		
9.9		39		42.5	8.75		18.8				59.3		32.8		
9.9	4.8	28.2	30.4	39.8	44.1	12.5	22.4	3.6		14.2	61.8	6.9	36.3	7.8	2.6
0	5.0	25.2	9.4	41.8	41.3	9.8	12.5	6.4	10.3	12.3	50.2	7.3	39.7	7.4	3.1
11.4	5.4	25.7	10.2	62.9	45.2	11.5	14.4	6.7	9.2	7.2	52.2	8.2	41.4	7.4	2.4
0	4.6	27.3	10.4	74	60	17.9	13.6	7.7	16.3	19.3	49.4	9.0	38.1	3.0	2.6
0.6	5.1	25.6	11.2	34.8	60.1	15.7	13.7	8.2	18.5	18.9	45.7	9.4	45.8	4.3	3.5
1	4.8	23.9	12.5	36.6	61	13.1	12.1	8.5	27	22.5	46.2	10.4	62.4	2.9	3.0
0.7	5.6	26.8	19.4	40.3	49.1	15.6	11.4	12.2	28.7	18.7	52.9	9.1	59.9	4.5	4.4
0.4	5.6	23.2	19.9	64.1	45.6	11.3	12.2	12.2	29.5	17.4	59.7	8.9	56.4	6.9	5.3
3.7	4.8	27.8	20.3	71.9	49.9	12.6	9.8	13.7	30.3	18.2	51.2	21.2	48.1	8.0	4.2
0.9	4.5	31.9	20.9	74.7	40.9	12.8	9.3	11.9	33.3	19.0	53.0	9	43.7	7.5	4.4
2.1	5.1	35.0	21.5	65.0	49.7	14.2	9.5	10.9	42.9	18.0	54.3	9.7	52.3	9.4	6.6
2.4	3.9	42.0	11.9	62.9	50.0	21.3	10.2	11.5	43.2	21.5	49.0	12.6	43.7	9.5	7.3
6.3	4.8	32.8	14.0	68.8	47.8	20.1	9.6	9.9	45	20.5	47.6	11.4	47.2	9.6	8.0
6.1	6.5	39.2	12.0	18.2	51.4	19.2	8.5	10.3	49.9	22.3	43.7	8.2	43.5	9.4	9.9
5.7	12.2	34.1	13.0	20.9	54.0	14.7	8.1	10	52.8	20.5	45.7	11.3	48.5	7.1	11.5
13.7	14.3	33.1	14.2	29.5	50.5	15.1	9.2	10.3	54.3	22.9	39.4	10.9	55.8	7.5	11.1
15.8	13.4	36.4	9.3	27.4	49.1	15.5	13.5	11.2	55.1	21.0	44.4	18.1	46	10.6	8.3
22.0	9.8	36.6	9.5	28.6	57.9	16.1	14.4	11.7	58.2	16.2	42.9	16.2	50.1	13.5	9.5
16.7	10.4	34.4	9.5	30.3	73.5	16.7	11.4	16.8	63.9	21.7	28.6	17.9	43.9	14.2	13.0
21.7	8.4	34.0	6.9	27.6	66.3	17.0	11.2	13.4	74.9	20.8	33.9	19.0	54.8	8.3	10.8
14.7	9.6	34.0	8.4	40.0	51.2	15.3	11.0	16.8	73.3	15.7	32.1	18.4	75.7	6.7	12.3
15.9	13.7	25.7	7.2	45.1	52.9	18.4	11.8	16.9	79.8	17.3	33.8	19.0	68.6	7.6	13.7
9.8	7.4	26.7	12.6	33.0	41.6	11.9	8.4	9.4	42.7	15.2	65.2	12.4	39.9	7.9	7.2
57	47	33	54	63	36	47	62	43	49	34	34	37	42	36	53
2.65	1.51	4.79	1.21	4.99	8.78	2.42	1.74	2.20	10.18	2.67	4.98	2.63	8.52	1.46	1.73
1.22	1.89	3.78	2.92	7.42	1.72	8.76	0.43	2.46	2.41	0.66	0.31	0.85	3.50	4.80	2.19

2.2.1 America

2.2.1.1 USA

General overview

While the USA is one of the few countries with reasonably detailed information on elasmobranch fisheries no comprehensive account of these fisheries on a national basis could be located. Main fisheries for elasmobranchs in the USA have traditionally been centred on sharks, although batoids have also been fished. Rays and skates were recorded in commercial catches as early as 1916 (Martin and Zorzi, 1993) mainly as by catch of more important fisheries. However, the first directed fisheries for elasmobranchs in the USA seem to have been for the tope shark, *Galeorhinus galeus* (then *zyopterus*), in California and for large sharks off Salerno in Florida. Both flourished as a consequence of the high demand for shark liver oil in the 1940s-50s and stopped mainly because of laboratory synthesis of vitamin A in 1950.

According to FAO statistics, until recently, the commercial catches of elasmobranchs in the USA were, together with those of Argentina, the least important among major elasmobranch-fishing countries in America. However, this has changed since the early 1990s. Elasmobranch production has varied considerably for the last 40 years oscillating around 10 000t/yr until the late 80's. Two periods of very low catches were 1952-1956 and 1970-1977, while 1958-1960 saw some of the highest yields. Since 1988 the post-war peak of 17 000t has been exceeded (Figure 2.2). Catches rapidly increased during the mid 1970s and soared in the mid-80's. Still, elasmobranchs are only a minor fishery as catches during 1987-1991 averaged only 0.42 % of the total fisheries production of the USA while representing 3.57% of the total reported elasmobranch catch in the world (Table 2.2).

According to Compagno (1990), the recent rise in catches might reflect a change in consumer preference that has made shark meat fashionable and acceptable to the public as a direct result of the infamous "Jaws" films. This would have prompted a whole new group of fisheries directed to sharks in the USA. According to Cook (1990), very recent changes in international shark-fin markets have further increased the demand for sharks in the USA. Amongst these new fisheries, those for the thresher shark, *Alopias vulpinus,* the Pacific angelshark *Squatina californica* and the shortfin mako *Isurus oxyrinchus,* are the most important in the West Coast. For the Gulf of Mexico and East Coast of the U.S.A, most of the recently increasing shark fisheries take a diverse catch of coastal sharks, reported as unclassified sharks. This difference in detail of the reported catches between the two coasts of the USA is probably because on the west coast there are different markets and prices for many species of elasmobranchs whereas on the east coast (NOAA 1991) only mako sharks attain a price different from the remaining "unclassified sharks."

Data from FAO shows that until 1980 elasmobranch catches in the USA were about evenly distributed on both coasts. Since 1981, the east coast has contributed the bulk of the catches as a result of a large expansion of fisheries for sharks and rays (Figure 2.3). This new growth led to the recent implementation management of large shark fisheries in the east coast. Overall, the two most important elasmobranch groups in the fisheries of the USA are the dogfishes (mainly *Squalus acanthias*) and the skates. Dogfish and skate catches from the waters within FAO Area 21 (roughly corresponding to the New England and Mid-Atlantic regions of the National Marine Fisheries Service of the USA) and dogfish catches in FAO Area 67 (roughly corresponding to the coasts of Washington and Oregon) have dominated the elasmobranch production of the country until recently.

Dogfish catches from the Northeast USA (Area 21) were the major part of total elasmobranch catches during 1979-1983, fell in 1984 and have slowly recovered since 1985. Skate catches in this region have increased tremendously since 1983. This made them the second most important group in 1989 with almost one third of the total elasmobranch catches of the country (Figures 2.3 and 2.4). Dogfish catches off the northwest USA (taken mainly in Washington) had fairly variable yields, and declining during the mid 80's, partially recovered in 1986-1987 only to subsequently fall. Most of the dogfish on the east coast and skates on both sides of the country are taken by trawlers while dogfish in the northwest coast are apparently harvested with gillnets and trawl nets. Although both rays and dogfish are low priced resources when compared with some other elasmobranchs (eg, mako or thresher sharks) they are available in such large quantities that they become profitable for fishing companies. There are apparently no management regimes specifically directed at the dogfish and ray resources of the USA. At most, some stocks are included in general management schemes for ground fish resources. Grulich and DuPaul (1987) estimate that the piked dogfish stocks of the US east coast could support a harvest of about 24 000t/yr in the mid-80's. However, recent studies suggest that the biomass of the *Squalus acanthias* stock sustaining most of this fishery, although increasing recently, is highly variable from year to year (Silva 1993). This could mean that high levels of exploitation are not sustainable and consequently supplies for a large market would be unreliable.

The East Coast

Throughout this century, the single most important fishery for sharks in the East Coast of USA was that for large sharks of Salerno Florida during the period 1935-1950 (major accounts are given in Springer 1951, 1960). The fishery depended on production of vitamin A from shark liver oil and failed when industrial synthesis of vitamin A began. The fins and hides were also utilized. The fleet was based at Salerno but during the summer it usually extended operations west to the Mississippi river and after 1945 expanded to include boats in the Carolinas, the Florida keys and the Gulf coast of Florida. The Caribbean and West Indies also provided catches to the company based at Salerno. In the later years approximately half of the catch came from the Gulf of Mexico. Up to 16 boats of 12-15.5m operated concurrently, fishing with two bottom longlines of at least 200 hooks in depths up to 90m. Floating longlines and bottom gillnets were also occasionally used. Sandbar sharks, *Carcharhinus plumbeus,* composed most of the catches, which peaked at 10 514 sharks in 1947.

In recent times, the second most important elasmobranch fisheries in the USA after dogfish and rays have been the growing fisheries for large sharks in the Gulf of Mexico and South Atlantic. While catches of large sharks have remained practically unchanged in the Mid-Atlantic and New England regions, shark catches in the Gulf of Mexico and South Atlantic regions underwent radical changes with an eightfold increase in yield from 1984 to 1989 (Figure 2.4). This trend, caused mainly by the development of a stable market, began in 1985 when fishermen began to target sharks with gillnets and longlines. The landing of previously discarded shark by catches from other fisheries also became profitable.

According to NOAA (1991), directed fisheries for sharks in the east coast include: a monofilament 18-64cm mesh driftnet fishery apparently targeted on schooling blacktip sharks in Florida; a May-November gillnet fishery in the east coast of Florida catching mostly *Carcharhinus* spp.; a driftnet fishery for tunas, billfishes and sharks in the Atlantic, Gulf of Mexico and Caribbean; pelagic longlines for tunas, billfishes and sharks in the Atlantic, Caribbean and Gulf of Mexico (this fishery deploys gear in a mechanized operation involving large vessels and thousands of hooks); a recent fishery for sharks with bottom longlines sets manually with up to 100 hooks from each small boat; and a pelagic hook and line fishery for tunas, billfishes and sharks in the Gulf of Maine, South New England and the Mid Atlantic.

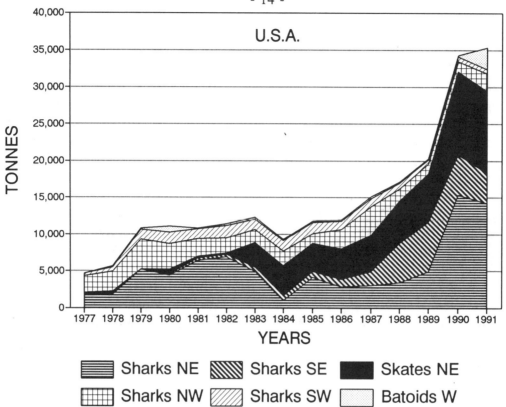

Figure 2.3. Elasmobranch catches of the USA by major groups and regions as reported by FAO during 1977-1991.

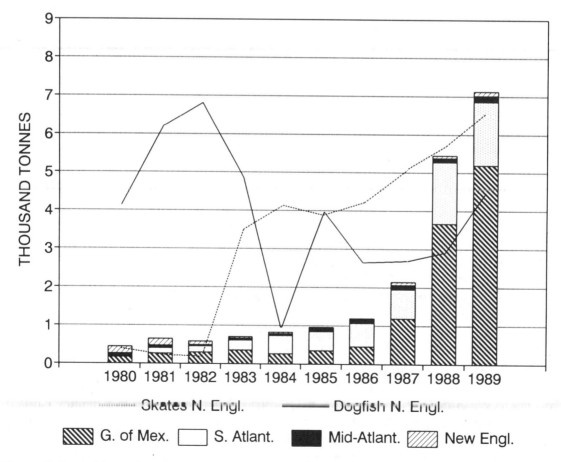

Figure 2.4. Elasmobranch catches from the east coast of the USA during 1980-1989. Bars represent shark fisheries. (Data from FAO and Hoff 1990).

Lawlor and Cook (1987) report that the seasonal East Florida longline fishery for sharks is carried out from boats 11-15.5 m long with 2-4 fishermen using bottom and/or surface longlines for periods of 1-2 days. The mainline varies from 1.6 to 10km in length and is made of 4.8-6.4 hard-lay tarred nylon, from which 300-500 ganglions of 3.6 m long multistrand steel cable fall, with 3/0 or 3.5/0 shark hooks each. Buoys are attached to the mainline on 28-30 m leaders for bottom longlines and for pelagic longlines with 10-30 m leaders. Bluefish, bonito, mackerel, mullet and squid are the most common bait. Apparently, about 110 boats work full-time and year-round in this fishery following migrating sharks along the coast. NOAA (1991) indicate that 124 vessels target sharks in the US east coast with longliner catches during 1989 adding up to 6140t while gillnetters caught 621t.

Some sharks in the east coast of USA are also landed as by catch from the following fisheries: the Gulf of Mexico tuna fisheries; the Gulf of Mexico and south Atlantic coast snapper-grouper bottom longline fishery; swordfish gillnet fishery of Massachussets and Rhode Island (up to 15 vessels) and the gillnet fisheries of Maine, Virginia, New York and New Jersey. The main species caught in the South Atlantic and Gulf of Mexico with gillnets are *Carcharhinus plumbeus, C. limbatus, C. leucas, C. altimus, C. brevipinna, Galeocerdo cuvier, Carcharias taurus, Negaprion brevirostris, Sphyrna lewini* and *S. mokarran*. Those captured with longlines are mainly *C. plumbeus, C. limbatus, C. isodon, C. acronotus, C. leucas, C. brevipinna, C. obscurus, Rhizoprionodon terraenovae, Carcharias taurus* and *Sphyrna lewini* (Hoff 1990; NOAA 1991). Russell (1993) reports *C. limbatus, Mustelus canis* and *Rhizoprionodon terraenovae* as the most common species caught by shark longliners in the northern Gulf of Mexico. Data from NOAA (1991) shows that ex-vessel prices for sharks in the Gulf of Mexico and southeast USA almost doubled from an average price in constant $US of $0.57/kg in 1979 to $1.12/kg in 1986, the average since 1983 being approximately $1.00/kg. Meanwhile, the prices for fins have risen nearly an order of magnitude since 1985. In general, higher prices are paid for dressed carcasses and for sharks fished in waters more than 3 miles from the coast as opposed to those caught inside the 3-mile state waters limit. The mako shark attains a higher price than the rest of the species which are treated as "unclassified shark."

Hoff (1990) stresses that important by catches of several species of sharks are taken regularly by the shrimp trawl fisheries in the northern Gulf of Mexico. Unfortunately, most of the catch is discarded as there is no market (GMFMC 1980). NOAA (1991) estimate that the incidental catch of sharks in the Gulf of Mexico shrimp fishery is of about 2800t/yr. Most individuals are juveniles from nursery areas and this catch might represent an important threat for recruitment to future breeding stocks. Escapement of larger specimens will probably increase if the regulations for the mandatory use of turtle excluder devices (TED) are approved. Overall, total yearly discards of sharks in all fisheries of the east coast of USA averaged 16 000t (NOAA 1991).

The great increase in shark exploitation both by commercial and recreational fishermen on the east coast of the USA led to catch quotas and bag limits in April 1993. This management took 10+ years to implement due to, among other things, lack of appropriate data for assessment regarding abundance, biology, distribution, life history and catches of shark. Given concerns about possible overexploitation of shark stocks during the late 80's, an assessment was performed with the available information. The estimated levels of long-term production are about 3400t for large coastal sharks and about 3600t for small coastal sharks (Parrack 1990, NOAA 1991). The species considered in each of the management units are listed in Table 2.3. A number of management measures aimed at rebuilding stocks in effect since April 1993 include: 1993 commercial quotas (in dressed weights) of 2436t for large coastal species and 580t for pelagic species; recreational bag limits of four sharks/vessel/trip for large coastal and pelagic sharks combined and five sharks/person/day for small coastal species; commercial fishing only by permit; fins landed in proportion to carcasses; release of shark by catches ensuring maximum probability of survival; compulsory submission of sales receipts and logbooks from selected commercial and recreational operators; presence of observers in selected commercial boats; and banning of shark catches for foreign vessels in US waters (NMFS 1993).

The West Coast

Holts (1988) and Cailliet et al. (1993) review the shark fisheries of the west coast of the USA. Aside from the piked dogfish fisheries which dominate the catches, an important group of directed fisheries for sharks suddenly arose in California at the end of the 70's. but some of have declined during the following decade. These fisheries arose mainly as a response to changes of trends in consumer preference which increased demand and prices for some species. Total catches (excluding dogfish) increased through the late 70's to a peak of about 1 800t in 1982 but have since varied with a decreasing trend (Table 2.4). Cailliet et al. (1993) consider market fluctuations and susceptibility to overexploitation of some stocks as the main reasons for diminishing catches.

The first species whose landings increased was the thresher shark *(Alopias vulpinus)* fishery centred between San Diego and Cape Mendocino. Operations started with 15 large-mesh driftnet vessels in 1977. Ex-vessel prices for this species increased from US$0.64/kg in 1977 to US$3.52/kg in 1986. The thresher shark fishery was soon displaced by the more valuable swordfish fishery and the thresher shark. This lead to social problems and poor management of the fishery and resulted in the loss of the thresher populations (see Bedford 1987 for a detailed account). Catches peaked in 1982 at 1083t when more than 200 vessels were operating, but slowly declined until 1986 when limited area and season legislation was passed. Catches further declined as a result of these regulations until the directed fishery for this species was banned in October 1990. At present only incidental catches are permitted which and they account for almost 300t/y (pers. comm., Holts, NMFS Southwest Fisheries Center) in the swordfish fishery. Throughout most of the fishery catches were composed mainly of young sharks 1-2 years old a few *A. superciliosus* and *A. pelagicus* are also included. Bedford (1987) reports that market sampling data showed decreasing modal sizes with time along with declining CPUE indices since the mid-80's. Unpublished data (Holts, pers. comm. op. cit.) shows the mean length of fish caught clearly declined during the same period.

Another recent development on the west coast was the fishery for Pacific angelshark *(Squatina californica)*. This began as a localized operation in Santa Barbara in 1977 (166kg landed), underwent a great expansion in 1981 (158t landed), reached a peak in 1986 (563t landed) and steadily declined in the following three years (121t in 1989, Table 2.4; Cailliet et al. 1993). Ex-vessel prices climbed from US$0.33/kg in 1978 to US$0.99/kg in 1984 (Holts 1988). Pacific angelsharks were taken initially as by catch of the Pacific halibut fishery with bottom set trammel nets. As markets and demand expanded, they began to be targeted with single-walled nylon twine (No. 24 to No.30) gillnets, 366-549m long and 13 meshes deep (mesh sizes between 30.5 and 40.6cm) (Richards 1987). Vessels were usually from the halibut fishery and used hydraulic gear retrievers. Operations were centred in the Santa Barbara-Ventura region and the Channel Islands in waters less than 20m deep, less than 1.6km offshore. In the opinion of Cailliet et al. (1993) the drop in catches since 1986 is due to a combination of declining availability of the species and changes in the market as cheaper imports of shark meat became available. The only regulations applied to this fishery are those pertaining to the set-net fishery for halibut in California. These neglect the need for separate management of the elasmobranch resources.

A shortfin mako *(Isurus oxyrinchus)* fishery in California also started as a valuable by catch of the driftnet fishery for swordfish and thresher shark in the late 70's. Catches increased

Table 2.3. Sharks species considered in each of the USA east coast management units (from NOAA 1991).

	FAO Common Name	Scientific Name
Large Coastal Sharks	Sandbar	*Carcharhinus plumbeus*
	Blacktip	*Carcharhinus limbatus*
	Dusky	*Carcharhinus obscurus*
	Spinner	*Carcharhinus brevipinna*
	Silky	*Carcharhinus falciformis*
	Bull	*Carcharhinus leucas*
	Bignose	*Carcharhinus altimus*
	Copper	*Carcharhinus brachyurus*
	Galapagos	*Carcharhinus galapagensis*
	Night	*Carcharhinus signatus*
	Caribbean reef	*Carcharhinus perezi*
	Tiger	*Galeocerdo cuvier*
	Lemon	*Negaprion brevirostris*
	Sandtiger	*Carcharias taurus*
	Bigeye sand tiger	*Odontaspis noronhai*
	Nurse	*Ginglymostoma cirratum*
	Scalloped hammerhead	*Sphyrna lewini*
	Great hammerhead	*Sphyrna mokarran*
	Smooth hammerhead	*Sphyrna zygaena*
	Whale	*Rhincodon typus*
	Basking	*Cetorhinus maximus*
	Great White	*Carcharodon carcharias*
Small Coastal Sharks	Atlantic sharpnose	*Rhizoprionodon terraenovae*
	Caribbean sharpnose	*Rhizoprionodon porosus*
	Finetooth	*Carcharhinus isodon*
	Blacknose	*Carcharhinus acronotus*
	Smalltail	*Carcharhinus porosus*
	Bonnethead	*Sphyrna tiburo*
	Sand devil	*Squatina dumeril*
Pelagic Sharks	Shortfin mako	*Isurus oxyrinchus*
	Longfin mako	*Isurus paucus*
	Porbeagle	*Lamna nasus*
	Thresher	*Alopias vulpinus*
	Bigeye thresher	*Alopias superciliosus*
	Blue	*Prionace glauca*
	Oceanic whitetip	*Carcharhinus longimanus*
	Sharpnose sevengill	*Heptranchias perlo*
	Bluntnose sixgill	*Hexanchus griseus*
	Bigeye sixgill	*Hexanchus vitulus*

Table 2.4. Shark landings, in dressed weight (kg), west coast USA (adapted from Cailliet et al. 1993)

FAO NAME	1976	1977	1978	1979	1980	1981	1982	1983	1984	1985	1986	1987	1988	1989
PIKED DOGFISH *	2,647,724	2,639,478	2,940,474	4,341,382	3,242,281	2,193,992	2,085,296	2,450,906	3,474,589	2,512,781	2,339,761	3,695,570	3,404,150	2,952,019
THRESHER SHARK	16	58,803	137,141	333,963	819,925	879,679	1,083,510	797,838	754,814	694,060	551,685	349,622	290,457	297,405
PACIFIC ANGELSHARK	313	166	37,402	56,138	49,950	118,054	144,351	159,510	287,496	561,966	563,473	426,845	223,072	121,786
SHORTFIN MAKO	9	9,040	12,456	16,042	70,523	125,227	239,585	146,621	110,786	97,667	207,053	277,857	222,105	176,298
TOPE SHARK	82,805	73,623	79,936	100,715	87,222	116,836	113,078	79,974	253,459	110,622	89,512	103,371	66,554	77,559
BLUE SHARK	1,041	44,658	16,300	38,121	87,227	92,116	26,258	6,348	1,789	1,070	1,294	1,774	3,301	6,184
LEOPARD SHARK	0	10,109	15,870	12,243	18,199	22,419	32,082	45,994	31,411	34,366	29,885	25,138	18,949	22,913
BIGEYE THRESHER SHARK	4,507	0	0	0	4,922	4,786	16,466	48,354	33,945	54,313	20,968	11,488	5,463	10,030
BROWN SMOOTH-HOUND	21,287	120	3,344	1,108	2,625	10,733	2,389	6,402	3,673	15,124	6,132	5,864	7,047	4,979
SMOOTH HAMMERHEAD	0	844	465	138	0	1,026	847	20,194	3,101	1,780	1,847	820	244	72
GREY SMOOTH-HOUND	27	0	15,320	5,469	345	0	1,144	479	3,108	851	230	0	9	187
HORN SHARK	6,624	525	124	9,559	3,843	1,038	3,424	220	278	165	89	24	62	15
GREAT WHITE SHARK	0	0	0	1,030	754	19	3,656	288	2,770	1,299	419	610	997	596
PELAGIC THRESHER SHARK	0	0	0	0	0	0	0	4,959	0	291	108	1,041	350	113
BROADNOSE SEVENGILL SHARK	0	0	38	0	247	1,550	927	788	128	405	25	77	10	6
COW SHARKS	23	0	113	132	199	350	603	571	605	194	199	163	71	59
SALMON SHARK	0	35	0	0	0	0	452	104	0	915	1,022	116	122	159
SWELLSHARK	0	0	1,269	0	74	0	0	0	101	9	0	0	1	2
BLUNTNOSE SIXGILL SHARK	0	0	0	9	5	144	0	58	44	2	0	25	9	61
DUSKY SHARK	0	92	47	0	0	89	0	54	0	0	0	0	0	0
UNSPECIFIED SHARKS	264,432	255,775	272,615	381,794	525,831	263,743	124,269	82,343	82,343	87,766	61,356	80,512	21,188	13,631
TOTAL	3,028,809	3,093,269	3,532,914	5,297,842	4,914,174	3,831,799	3,878,318	3,852,008	5,044,440	4,175,643	3,874,858	4,980,917	4,264,160	3,684,074
TOTAL EXCLUDING DOGFISH	381,084	453,790	592,440	956,460	1,671,892	1,637,808	1,793,021	1,401,102	1,569,852	1,662,863	1,535,097	1,285,347	860,009	732,055

* includes catches from Canadian waters (approx. 50% during 1983-89)

steadily from 1977 through 1982 when they reached 239t then underwent a period of lower levels possibly owing to changes in fishing strategy or environmental conditions (Holts 1988) but peaked again in 1987 at 277t. Since then, catches have declined once more (Table 2.4). The bycatch of makos in the driftnet fishery is low and since 1988 a closely controlled experimental fishery was started with longlines targeting this species. Under this regime, 6 vessels using 4.8-8.2km stainless steel cable longlines near the surface are allowed to fish subject to time/area closures and away from sport fishing grounds. Additionally, a TAC has been established at 80t and a market for the substantial blue shark by catch must be developed to utilize this resource. By catches of shortfin mako in the driftnet fishery are also allowed. Although the shortfin mako fishery is mainly sustained by very young sharks averaging 9-14kg dressed weight, there is no apparent decline in the mean size of the catches. Populations look healthy and even might be relatively lightly exploited (Holts 1988, Cailliet et al. 1993).

In addition to these three fisheries which constitute the main "new" shark fisheries in the last 15 years on the west coast, many other elasmobranchs are also taken commercially, mainly as a by catch of other fisheries. Martin and Zorzi (1993) review the skate fisheries of California. Skates (mainly *Raja binoculata, R. inornata* and *R. rhina)* have been fished in California since at least 1916, averaging 96t/yr and 11.8% of total commercial elasmobranch catches in California. San Francisco and Monterey are the main landing ports receiving 70% of the total. There are technical constraints in the processing marketable skates of sizes up to one kilogram and most of the landings of *R. binoculata* and *R. rhina* consist of immature individuals. Roedel and Ripley (1950) suggest that the skate resource might be underutilized, but it also seems to be presently misutilized. A market for larger skates should be developed if this resource is to be properly used and managed.

Another species of interest is the blue shark *(Prionace glauca)*. Holts (1988) and Cailliet et al. (1993) summarize the available information. The blue shark is a major incidental catch of the driftnet fishery of California and a minor by catch of the set-net fisheries for halibut and angel sharks. Mortality estimates for the driftnet fishery were 15 000-20 000 (300t) sharks annually in the early period, although changes in gear design have reduced this mortality. The experimental longline fishery for mako sharks also takes incidental catches of blue sharks at a rate of four blue sharks for each mako. Nevertheless, a conservation programme of enforced rapid release of live sharks is expected to decrease this mortality. A small experimental longline fishery with one vessel occurred during 1980-1982 and catches of blue sharks peaked around 90t in 1980 and 1981 (Table 2.4). The main constraint for the development of a large scale fishery for blue sharks is the lack of markets. Blue shark meat is reportedly less palatable than that of other clasmobranchs. Attempts to start a fishery for salmon sharks *Lamna ditropis* in Alaskan waters was reported (Paust 1987) but no other records were found.

The single most important fishery for elasmobranchs off the west coast of the USA was that developed in California for tope shark *Galeorhinus galeus* during the 1930's-1940's. Ripley (1946) gives a detailed description of this fishery. Stimulated by the discovery in 1937 that the tope sharks of that area were the richest source of high potency vitamin A in the world, the subsequent 4 years saw increases in catches that reached over eight times those of pre-boom levels and averaged approximately 3400t/yr. Vessels from the northern halibut fishery switched to shark fishing and in a short period all sorts of vessels modified their operations and joined the fishery totalling about 600 boats by 1939. Swift changes in gears from drift and set gillnets to machine-handled halibut longlines and back to "diver" gillnets and the posterior mechanization of their operation occurred in a period of less than 3 years (detailed description of gears used are given in Roedel and Ripley 1950). Northern California was the main fishing area with more than 70% of the catches although fishing occurred along the entire coast, mostly within 7.8 km of shore in waters up to 144m deep. After 1941, catches plummeted and never recovered their former levels. The discovery of synthetic vitamin A prevented efforts to revive this fishery, although a small fishery has continued to present times. Catches since 1976 fluctuated between 66 and 253t/yr (Table 2.3). Activities are now centred around San Diego and Orange counties (Holts 1988) apparently as an incidental consequence of net fisheries for halibut and angel shark. Only general regulations for the latter fisheries "protect" tope shark populations. Holden (1977) estimates the north Pacific unexploited stock size at 29 400t, but it appears that stocks have not yet recovered to former levels (Holts 1988). However, no recent assessments have been done for this species. Finally, a short lived small-scale harpoon fishery for basking sharks (*Cetorhinus maximus*) existed during the late 40's in Pismo Beach (Roedel and Ripley 1950) but also ceased as a consequence of the fall of the liver oil industry.

2.2.1.2 Mexico

Since the mid-70's, Mexican elasmobranch fisheries have been the largest in America (Figure 2.2). FAO statistics show that there has been a general trend of increased catches of elasmobranchs in Mexico, from the typical 5 000t/yr of the 50's to recent levels of varying around 30 000t/yr since the early 80's. Judging from the trend of the last ten years, Mexican fisheries for sharks and rays have attained relative stability. Elasmobranchs are a relatively important resource in Mexico, comprising 2.36% of the national catches during 1987-1991. This figure is comparable with other major elasmobranch-fishing countries but is substantially higher than the 0.8% contribution of elasmobranchs to world fisheries in the last 10 years. Elasmobranch exploitation in Mexico can be traced back to at least the 1930's, but detailed statistics are difficult to find before the mid-1970s. Walford (1935) reports "several tons" of shark fins from the west coast of Mexico being imported to California each year and Ripley (1946) refers to Mexican fisheries supplying shark liver oil to the USA industry. Mazatlan and Guaymas were the main ports in the west coast shark fishery. Catches peaked at 9 000t in 1944 but declined to 480t in 1953 after the fall of the shark liver oil industry (Castillo 1990). On the east coast during the 40's, a fleet based at Progreso, Yucatan targeted sharks had characteristics similar to the fleet of Salerno, Florida, and caught up to 3200t/yr since 1950 (GMFMC 1980).

Mexican fisheries for elasmobranchs are targeted on sharks. Batoids are seldom exploited but considerable (and unknown) amounts are discarded in the extensive trawling operations for shrimp fisheries. According to data from the Mexican Ministry of Fisheries yearbooks for 1977-1991, sharks account for 94.8% (29 036t/yr) of elasmobranch catches while batoids only represent 4.2% (1272t/yr).

Because of its larger coastal extension, the Pacific coast contributes 60% of total shark catches while the remaining 40% comes from the Gulf of Mexico and Caribbean. No data on catches by species are available. Only small sharks (those measuring less than 1.5m total length (TL) when caught and are know locally as cazón) and large sharks (those larger than 1.5m TL) are recorded in the statistics. Large sharks comprise 60% of total shark catches and ⅔ of these are caught in the Pacific while only ⅓ are caught in the Gulf of Mexico and Caribbean. The remaining 40% of the total shark catches are small sharks, 64% come from the Pacific and 36% from the east coasts. There is some variability in the catches of large and small sharks from each coast, but overall, Mexican fisheries seem to have reached an equilibrium during the last 10 years (Figure 2.5). Meanwhile, batoid catches are slowly and steadily expanding.

Mexican shark fisheries are largely artisanal, multispecies, multigear fisheries. Bonfil et al. (1990), Castillo (1990) and Bonfil (*in press*) summarize most of the available information on elasmobranch fisheries in Mexico. They estimate that about ⅓ of the shark catch is taken by small-scale fisheries. Vessels are generally made of fibreglass, 7-9 m long with outboard motors using either gillnets or longlines depending on the regional customs. Some vessels of 14-20 m are also used whereas only a few vessels in excess of 20 m take part in the fishery. Significant quantities of sharks and rays are also taken as incidental catches of large-scale trawl fisheries for shrimp or demersal fishes in both coasts. Large scale fisheries for tunas and billfishes in both coasts also contribute to the total catches. Sharks and rays are traditionally used for food in Mexico, either fresh, frozen or more commonly, salt-dried. Shark fins and hides are also exported and most offal is reduced to fish meal.

The main fishing grounds in the Pacific are centred in the Gulf of California in the north and the Gulf of Tehuantepec in the south. However, most of the available information about these fisheries comes from the northern area. Apart from the total catch, little is know about the shark fisheries in the Gulf of Tehuantepec. In the northern part, sharks are mainly caught with monofilament longlines of 1-2 km and approximately 350 hooks, although smaller quantities are taken with gillnets of up to 2 km long. Some 17 vessels, 44 m long and using longlines of up to 2000 hooks targeted sharks and billfishes on the Pacific coast during 1987. It is unknown if these vessels are still operating. A similar number of Japanese-Mexican joint venture longliners caught 234t/yr of sharks in Baja California during 1981-1983 (Holts 1988).

Fishing grounds span the entire east coast. During 1976-1988, Veracruz and Campeche shared 58% of the total shark catch and Tamaulipas and Yucatán 30%. Longlines are utilized mostly in the state of Veracruz and presumably also in Tamaulipas. Gillnets from 11-40cm mesh size are the main fishing gear in the Bank of Campeche. There is a substantial by catch of mainly juvenile sharks in the semi-industrialized longline fisheries for red grouper and red snapper on the Campeche Bank but no estimates of this catch are available.

Information about the species caught in the different regions of the Mexican coast and the composition of the catches is incomplete. Most of the available research has been done in the mouth of the Gulf of California on the west coast and in the southern States of Campeche, Yucatan and Quintana Roo on the east coast. Important landings also occur in other areas of both coasts but have been poorly documented.

At least 44 species of shark are reported in the commercial catches of Mexico and 12 are the most important in the catches in the area of La Paz, Baja California and Sinaloa whereas 15 are the main species in the Gulf of Mexico and Caribbean (Table 2.5). Most of the large sharks caught consist of *Carcharhinus* spp, *Sphyrna* spp and other carcharhinids, while the small shark catches are a mixture of mainly *Mustelus* spp. and *Rhizoprionodon* spp., with juveniles of the large sharks sometimes contributing an important part of the total. Along the Sinaloa coast in the central Pacific *Rhizoprionodon longurio*, *Sphyrna lewini*, *Nasolamia velox*, *Carcharhinus limbatus*, *C. falciformis*, *C. leucas* and *Galeocerdo cuvier*, are the most important species. Galván-Magaña et al. (1989) report that *Mustelus lunulatus*, *Heterodontus mexicanus* and *Sphyrna lewini* are the most important sharks in the area of La Paz, B.C.. Experimental catches of longliners in the Pacific caught mainly pre-adult and adult *Alopias vulpinus* and *Carcharhinus limbatus* (Velez et al. 1989). For the east coast, the most important species are *Carcharhinus falciformis*, *C. leucas*, *C. obscurus*, *C. plumbeus*, *C. limbatus*, *Rhizoprionodon terraenovae*, *Sphyrna tiburo*, *Mustelus canis*, *C. brevipinna*, *Negaprion brevirostris*, *Sphyrna mokarran*, *Sphyrna lewini*, *Galeocerdo cuvier* and *Ginglymostoma cirratum*. With the exceptions of *C. obscurus* and *Ginglymostoma cirratum*, all the important species of the east coast are known to be heavily exploited as juveniles and sometimes even as newborns, at least in some part of their range.

Table 2.5.   Shark species found in the commercial fisheries of Mexico.

FAMILY		SPECIES	PACIFIC	GULF OF MEXICO /CARIBBEAN
Hexanchidae	1	*Heptranchias perlo*		X
	2	*Hexanchus griseus*		X
	3	*Hexanchus vitulus*		X
Echinorhinidae	4	*Echinorhinus cookei*	X	
Squalidae	5	*Centrophorus granulosus*		X
	6	*Centrophorus uyato*		X
	7	*Squalus cubensis*		X
	8	*Squalus mitsukurii*		X
Squatinidae	9	*Squatina californica*	X*	
Heterodontidae	10	*Heterodontus mexicanus*	X*	
Ginglymostomatidae	11	*Ginglymostoma cirratum*	X	X*
Rhiniodontidae	12	*Rhiniodon typus*	X	X
Alopiidae	13	*Alopias vulpinus*	X*	
	14	*Alopias superciliosus*	X	X
Lamnidae	15	*Isurus oxyrinchus*	X	X
Triakidae	16	*Mustelus californicus*	X	
	17	*Mustelus canis*		X*
	18	*Mustelus lunulatus*	X*	
	19	*Mustelus sp. ?*		X
	20	*Triakis semifasciata*	X	
Carcharhinidae	21	*Carcharhinus acronotus*		X*
	22	*Carcharhinus altimus*	X	X
	23	*Carcharhinus brevipinna*		X*
	24	*Carcharhinus falciformis*	X*	X*
	25	*Carcharhinus leucas*	X*	X*
	26	*Carcharhinus limbatus*	X*	X*
	27	*Carcharhinus longimanus*		X
	28	*Carcharhinus obscurus*	X	X*
	29	*Carcharhinus perezi*		X
	30	*Carcharhinus plumbeus*		X*
	31	*Carcharhinus porosus*	X	X
	32	*Carcharhinus signatus*		X
	33	*Galeocerdo cuvier*	X*	X*
	34	*Nasolamia velox*	X*	
	35	*Negaprion acutidens*	X	
	36	*Negaprion brevirostris*		X*
	37	*Prionace glauca*	X*	
	38	*Rhizoprionodon longurio*	X*	
	39	*Rhizoprionodon terraenovae*		X*
Sphyrnidae	40	*Sphyrna lewini*	X*	X*
	41	*Sphyrna media*	X	
	42	*Sphyrna mokarran*	X	X*
	43	*Sphyrna tiburo*	X	X*
	44	*Sphyrna zygaena*	X	

* Main species in the commercial catches.

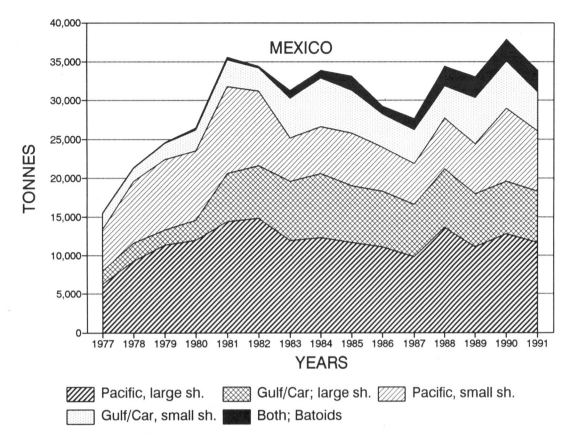

Figure 2.5. Elasmobranch catches in the Pacific and Gulf of Mexico/Caribbean coasts of Mexico during 1977-1991. (sh = sharks). (Data from Secretaría de Pesca, México).

A few isolated preliminary assessments of the status of some shark stocks exist for the east coast. Alvarez (1988) reports that surplus production models show that the stocks of *Sphyrna tiburo* and *Rhizoprionodon terraenovae* in Yucatan are close to optimal exploitation levels; results of the yield-per-recruit model suggest exploitation of *Sphyrna tiburo* is at the optimum level whereas *Rhizoprionodon terraenovae* seems to be already overexploited. For the production models, catch and effort were estimated in a very rough way and for the dynamic model, growth and mortality were estimated via length frequency analysis. Bonfil (1990), estimated growth via vertebrae readings and using the yield-per-recruit model found growth overfishing for the *Carcharhinus falciformis* stock of the Campeche Bank. This results mainly from the high catches of newborns and juveniles of this species in the local red grouper fishery.

There have been several permanent research programmes for shark fisheries in Mexico since the early 80's. Despite this, to date Mexico has no specific management for elasmobranch fisheries. A number of concerns have been expressed about undesirable practices in the fisheries. At least, *Carcharhinus falciformis, C. acronotus, Rhizoprionodon terraenovae* and *Sphyrna tiburo* are heavily exploited as juveniles in Campeche and Yucatan, hence raising the possibility of a future collapse of their stocks. Further, there are indications that large decreases in the abundance of juveniles of *C. leucas, C. limbatus, C. acronotus, C. perezi* and *Negaprion brevirostris* have occurred in some coastal lagoons of the Yucatan Peninsula as a direct consequence of heavy fishing with set nets (Bonfil *in press*). It is likely that this is commonplace in most coastal lagoons along the coast of Mexico. Further, the killing of large numbers of pregnant female *Rhizoprionodon longurio* in Sinaloa, on the west coast is another concern.

Although information is poor it is likely that many stocks in the Gulfs of California and Tehuantepec are close to their optimum level of exploitation or are even overfished. However, no assessments are known to date. Limited or non-existent information about the size of the stocks and about the actual levels of mortality makes an adequate appraisal of the status of Mexican shark fisheries difficult.

As in other countries, socio-economic and health problems related to the fisheries complicate the management of elasmobranchs in Mexico. The chances of curtailing the fishing of juvenile sharks in Mexico is constrained by the problems of the artisanal nature of many of the fishing fleets (loss of income for large numbers of fishermen) and the high demand for small sharks. The higher concentration of heavy metals generally found in older sharks also makes the harvesting of juveniles preferable.

2.2.1.3 Peru

From the mid-sixties until recently, the elasmobranch catches of Peru were the third largest in America and contributed 2.71% to the world elasmobranch catch. Nevertheless, elasmobranchs are of minor importance in Peru and represent only 0.29% of the total fishery production (Table 2.2). Their elasmobranch fisheries had a fairly steady trend of slow development in the 50's and early 60's. Since the mid-1960s catches have oscillated around 18 000t, peaking at more than 30 000t in 1984 and crashing in 1990-1991 (Figure 2.2). There may be a link between recently declining elasmobranch catches and the eruption of cholera in Peru during 1990.

Elasmobranch production in Peru is strongly dominated by smoothhounds. During the period 1977-1991, smooth-hounds of the genus *Mustelus* were the most important species in the elasmobranch catch making 5% (10 219t/yr) of the total and accounted for 25 000t in 1984 when record elasmobranch catches of 34 400t were taken, (Figure 2.6). Unspecified rays comprise 25% (4640t/yr) of the total catches. Their landings have increased significantly since 1984, making them the second most important elasmobranch group. *Rhinobatos planiceps* and angel sharks, *Squatina* spp., are also important species with average catches of 10% (1908t/yr) and 3% (560t/yr) respectively. The yields of these two groups showed variable trends in this period. An assorted group of elasmobranchs comprise the remaining 6% (1133t/yr). Apart from FAO statistics, nothing else is known about the elasmobranch fisheries of Peru.

2.2.1.4 Brazil

The Americas Brazilian elasmobranch catches follow those of Mexico and the USA, in size. It appears that Brazilian elasmobranch fisheries have attained a good degree of stability. After a slow but steady start through the sixties and a brief fall in the 70's, the catches of sharks and rays from Brazil underwent a major leap in the early 80's. Yields have since varied up to a maximum of 30 000t (Figure 2.2). Sharks and rays contributed 3% to the total catch during 1987-1991 making 4.0% of the world catches of elasmobranchs (Table 2.2).

Statistics do not differentiate elasmobranchs by species in Brazil. At least 30 elasmobranchs are common in the commercial catches in the southeast, but most of the landings are dressed and without head or fins making it difficult to distinguish species (Tomas 1987).

Some of the species mentioned in commercial catches are: Mustelus schmitti, Galeorhinus galeus, Prionace glauca, *Isurus oxyrinchus, Squatina guggenheim, Squatina* sp, *Pristis* spp., *Rhinobatos percellens, R. horkelii, Dasyatis* spp, *Gymnura* spp and *Myliobatis* spp.

According to FAO data Brazilian landings during the period 1977-1991 have been dominated by an assorted group of species corresponding to 72% (17 919t/yr) of the elasmobranch catches. Yields for this group of elasmobranchs grew rapidly from less than 1 000t in 1978 to more than 23 000t in 1982 and have remained close to 20 000t/yr since then (Figure 2.7). All the sharks known to occur in Brazilian catches are included in this group. According to Batista (1988) landings of *Galeorhinus galeus* have increased since 1970 due to increased trawling in south east Brazil. The second most important group during this period were the skates and rays comprising 17% (4254t/yr) of the catches. Landings of this group, as well as those of guitarfishes *Rhinobatos* spp. which averaged 7% (1683t/yr) of the total elasmobranch catch, expanded slowly. Small catches of sawfishes *Pristis* sp. have been steadily landed averaging 4% (1014t/yr) of the catch.

Vooren and Betito (1987) report on at least 25 species of small sharks and 24 of batoids found in waters less than 100m deep in the southeastern continental shelf. Swept area biomass estimates indicate that 20 000t are available in winter and 13 000t in summer. Of these 90% consist of 16 small sharks and 8 batoid species of commercial value. Apparently the only traditional use for elasmobranchs in Brazil has been for food, but Göcks (987) and Jacinto (1987) note some efforts to use the hides and other parts.

At least two kinds of fisheries land elasmobranchs in the north of Brazil (R. Lessa, pers. comm). An industrial longline fishery for tunas with up to 50% bycatches of sharks, takes mainly *Prionace glauca, Carcharhinus longimanus, Carcharhinus* spp., *Sphyrna* spp., *Isurus* spp. *Alopias* spp., *Pseudocarcharias kamoharai* and *Galeocerdo cuvier*. This fishery landed an average of 144t/yr of sharks between 1985-1990. About 60% of these were sharks less than 1.5m TL. Artisanal fisheries to catch *Cynoscium acoula* and *Scomberomorus* spp. catch *Carcharhinus porosus, Rhizoprionodon* spp., *Sphyrna* spp., *Isogomphodon oxyrhynchus* and *Pristis peroretti*. There is a high incidence of juveniles in this fishery which uses small driftnets about 1km long and 6m deep. Along the north shore between the Amazon river and Recife elasmobranch catches comprise up to 60% of the total catch. Incidental catches of small sharks and rays in the *Brachyplatystoma*, shrimp and snapper fisheries in the north of Brazil are reported by Evangelista (1987). Apparently most of the bycatches were formerly discarded but are now beginning to be used.

Vooren et al. (1990) summarize information on demersal fisheries for elasmobranchs during 1973-1986 on the continental shelf off the southern port of Rio Grande. Elasmobranchs account for 7.3% of the total catches, 13.1% of the trawl catches, 7.1% of the paired trawl catches and 5.4% of small-scale fisheries catches. Trawling is done with 440-480 HP boats of 11-13 day trips in depths between 40-100m while paired trawling is done by 340-370 HP boats of 9-11 day trips in depths less than 40m. Small-scale fisheries include beach seining and trammel nets used in waters less than 10m deep and gillnetting by 11-16m boats with 100-130 HP motors in waters 8-40m deep. Small sharks average 46.3% of elasmobranch catches while angel sharks, guitar fishes and rays account for 24.85%, 24.5% and 5%, respectively, of the catch. *Mustelus schmitti* and *Galeorhinus galeus* comprise most of the catches of "caçoes" or small sharks, and show increased landings, from 1414t in 1973 to 3217t in 1986, but, according to SUDEPE (1990), landings to 2023t in 1989. The proportion of small sharks in the catches of the small-scale and pair-trawler fishery increased during this period but decreased in the trawl fishery. This resulted in almost equal landings by each but fishery in 1983-1986. CPUE of

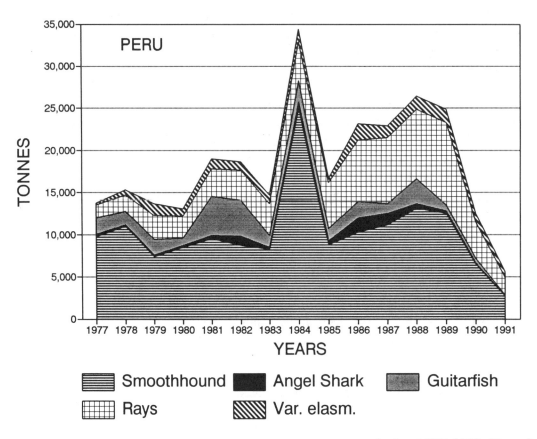

Figure 2.6. Elasmobranch catches of Peru, by species groups, during 1977-1991 (Data from FAO).

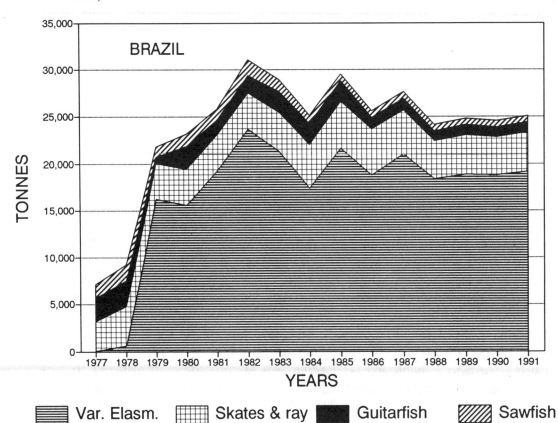

Figure 2.7. Elasmobranch catches of Brazil, by species groups, during 1977-1991 (Data from FAO).

small sharks for both types of trawlers tended to increase throughout the study period. Angel sharks (*Squatina guggenheim* and *Squatina* sp.) landings increased from 822t in 1973 to 1777t in 1986. As with the small sharks, the proportion of catches contributed by small-scale fisheries and pair trawlers increased while that of the trawl fishery decreased. Still, about 50% of the total landings of angel sharks came from the latter. While CPUE of angel sharks from trawlers showed an overall increase, paired trawlers' CPUE increased until 1983 and decreased afterwards. Landings of guitar fish, *Rhinobatos horkelii*, varied between 600 and 1925t. Most of this came from the small scale fisheries (50%) and paired trawlers (32%), while trawlers contributed very small catches (13%). Data of CPUE showed a slight decrease until 1982 for both types of trawlers, increasing to 1984 and then falling. Landings of rays, mainly of *Dasyatis spp.* and *Gymnura* spp., and to a lesser extent, *Myliobatis*, grew from 36t in 1973 to 484t in 1986. Small-scale fisheries averaged 18% of these catches, paired trawling 53% and otter trawling 34%. CPUE for rays in trawl fisheries were variable with an increasing trend.

The apparent decline of some of these populations in the last period of the above study seems to be confirmed by a switch from trawling to bottom longlines and gillnets (the latter specifically aimed at *Squatina* and *Galeorhinus*) which started in 1986 due to decreasing CPUE. This switch was coupled with additional fishing for angel sharks by shrimp trawlers from other areas during the off-season for shrimp (Pers. comm., C.M. Vooren, Universidad de Rio Grande, 1991).

Amorim and Arfelli (1987) and Arfelli et al. (1987) report some bycatches of large sharks in southern and southeastern waters by tuna longliners. *Prionace glauca* accounted for 33% of total catches of this fleet in 1985 and *Isurus oxyrinchus* accounted for 3.2% of total catches during 1971-1985. They are caught mainly during April-July and May-November. Landings of blue sharks consist mainly of carcasses of 20-40kg dressed weight (no head, fins or guts) which accounted for 553t and 462t in 1984 and 1985 respectively. Blue shark CPUE has varied from 0.4 kg/100 hooks in 1971 (when their capture was avoided) to 27.6 kg/100 hooks in 1985. Shortfin mako catches varied between 21t (1971) and 73t (1981), their mean weight in the catch varying between 42kg and 60kg throughout 1985. They are the most valued of elasmobranchs in Brazil and are consumed locally and exported to the USA.

Much research on elasmobranchs is done by Brazilian Universities, governmental and non-governmental organizations. However, according to Lessa (pers. comm., op.cit.), at present there are no management measures for elasmobranchs in Brazil although some local groups intend to raise governmental concern about the status of these fisheries. There are plans to report landings by species and her communication notes that elasmobranch stocks exploited by the north coast artisanal fishery are thought to be underexploited, those utilized by the tuna longline fishery are sustainably exploited and the south Brazil demersal stocks are overexploited.

2.2.1.5 Argentina

Elasmobranch catches of Argentina are one of the few expanding major elasmobranch-fishing countries in America. After a temporary drop in the late 40's, attributed to the collapse of shark liver oil fisheries, shark and ray production had a slow but steady growth from the early 1950s to the mid 1960s (Figure 2.2). Since 1967, yields have fluctuated around 10 000t and have increased since 1981. Despite the relatively low catches, which accounted only for 2.54% of the world elasmobranch catch during 1987-1991, elasmobranchs are reasonably important for Argentinean fisheries contributing 3.19% of the total yield during this period. This is the highest relative importance of elasmobranchs in major American elasmobranch fishing countries.

During 1977-1991 the most important species in the elasmobranch catches were: stat. *Mustelus schmitti* which averaged 49% (6790t/yr) of the total elasmobranch catch; several rays at 20% (2722t/yr), unclassified elasmobranchs at 23% (3160t/yr) and elephant fishes (*Callorhinchus* sp.) at 8% (1048t/yr). Of these groups, catches of smooth-hounds and "various elasmobranchs" had an increasing trend while elephant fishes and rays had a decreasing tendency. Argentina is one of the few countries in the world, with important catches of chimaeriformes (Figure 2.8).

Crespo and Corcuera (1990) give a detailed description of the fisheries for sharks off Claromeco and Necochea, Buenos Aires Province. In this northern Argentine fishery, gillnets are used to catch *Galeorhinus galeus*, *Mustelus schmitti*, *Carcharias taurus* and *Squatina argentina*. About 23 vessels, from 8-44.9m in length prosecute this fishery. They use nylon monofilament gillnets (2-3mm twine) with 19-21cm mesh, 55-71m long, 3.8m deep and 8-25 panels. These gillnets are set on the bottom between 0.5 and 25nm from the coast in depths from 2-70m. Usual catch per panel is 6-15 *Squatina argentina* and 1-20 of the other sharks species. Ex-vessel prices are US$3-4/kg for undamaged *Galeorhinus* destined for export (mainly to Italy) and US$1-2.5/kg for damaged ones that are consumed salt-dried in the local market. These authors report extensive damage to shark catches by marine mammals. Sea lions bite out the belly of entangled sharks and eat the liver.

Menni et al. (1986) note the presence of more sharks in the catch in northern Argentina. In addition to the species mentioned above, they report *Mustelus canis, M. fasciatus, Squalus blainvillei, S. cubensis* and *Notorhynchus cepedianus* in the commercial catches of Buenos Aires province. *Mustelus schmitti* accounted for 92% of their shark samples at commercial landing sites. The remaining species are less than 1% of the shark catch except *S. cubensis* which made up 2%. Government statistics of shark landings at Mar del Plata port averaged 5890t during 1971-1980. This is about ¥ of the average total elasmobranch catch of Argentina during that period. About 93% of this catch is made of 'gatuzos' (predominantly *Mustelus schmitti*, with some quantities of *M. canis* and some small numbers of *M. asciatus*). Cazones, (mainly *Galeorhinus galeus* but including some large *M. canis*) contributed the remaining 7%. Apparently, the remaining species are not recorded in the statistics.

2.2.2 Europe

2.2.2.1 Norway

Some of the most important shark fisheries in the North Atlantic have been carried out by Norwegian vessels. These fisheries have varied since the end of World War II with an increasing trend up to 1963, followed by a general decrease to levels around 7500t/yr since 1981 (Figure 2.2). Catches rose in the last three reported years. Elasmobranchs are not important for Norway judging from recent trends which show that elasmobranchs represented only 0.44% of the total fisheries production of Norway. Moreover, Norwegian shark and ray fisheries contribute only 1.21% to the world elasmobranch production during 1987-1991 (Table 2.2).

Catches of piked dogfish *Squalus acanthias*, have commonly accounted for the largest part of the total elasmobranch catches. Nevertheless, important fisheries for porbeagles existed in the 60's and for basking sharks during the last decade. While marketing and economical constraints have traditionally inhibited basking shark fisheries (Maxwell 1952; O'Connor 1953; Kunslik 1988), apparently the porbeagle, (*Lamna nasus*), fishery declined, at least in part, as a result of over-exploitation (Gauld 1989, Myklevoll 1989a, Anderson 1990).

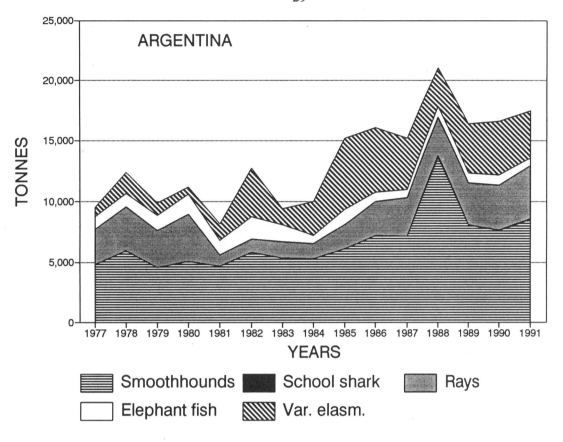

Figure 2.8. Elasmobranch catches of Argentina, by species groups, during 1977-1991 (Data from FAO).

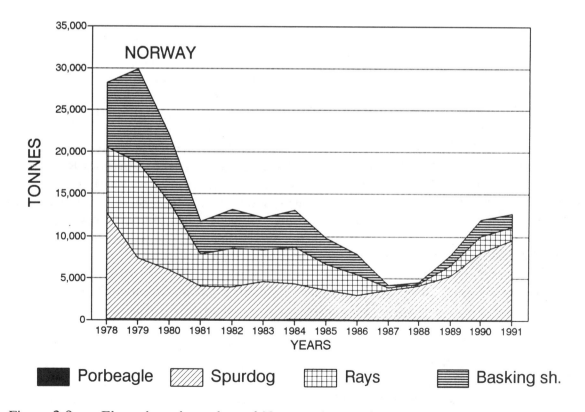

Figure 2.9. Elasmobranch catches of Norway, by species groups, during 1977-1991 (Data from FAO).

Norwegian elasmobranch fisheries are recovering after a prolonged decline. For the first time in almost 20 years, catch trends are increasing. FAO data for 1978-1991 (Figure 2.9) show catches of piked dogfish declining from more than 12 000t in 1978 to 2986t in 1986 then rising to 9627t in 1991, averaging 5715t/yr (53% of elasmobranch catches for this period). Catches of basking sharks,(*Cetorhinus maximus*) show a pattern similar to that of piked dogfish although their recovery is more modest. Basking shark catches fell from 11 335t in 1979 to only 352t in 1987, but were 1932t in 1990 and averaged 3929t/yr (36%) during this period. Catches of rays are fairly stable around 1115t/yr (10%). Small quantities of porbeagles are still caught on average 67t/yr.

Although the published data from the directed Norwegian fishery of the 60's is not considered (Gauld 1989; Anderson 1990), it is clear that this fishery caught large amounts of porbeagles. The summary of this fishery given here is based on Aasen (1963) and Myklevoll (1989a). Operations started as a coastal activity and after 1930 expanded from Norwegian waters northwest to the Orkney-Shetland area and the Faroes, then south into Irish waters and finally went to Canada and northern USA. Distant water operations by specialised freezer vessels 43-50m long deployed longlines with up to 5000 hooks in waters 10-30 m deep. Sharks less than 10kg were discarded as no market for them existed. The home fleet consisted of wooden boats 23-30m long which kept the catch on ice. Once the NW Atlantic porbeagle stocks declined to unprofitable levels by 1965, the fleet switched to mako sharks off North West Africa. Dressed carcasses of porbeagles were exported frozen to Italy while fins were marketed in the Far East. At present, only by catches of porbeagles from purse-seining, trawling and gillnet fisheries are landed. Norwegians do not even take their 200t TAC in EC waters.

The basking shark fishery started in the 16th century when the dried flesh was used as food (Kunzlik 1988 and Myklevoll 1989b), and has been an important tradition directed fishery. The major expansion of the fishery started in 1960, stimulated by demand for liver oil. Small wooden vessels 15-25m long, using harpoons operated mainly during April-August. Experiments to use the flesh of basking sharks (for fishmeal) and their hides failed. Consequently, in practice comparable to "finning", Norwegian fishermen took just the liver for oil extraction and discarded the carcasses. Later, they also took the fins for export to the Orient. During 1959-1980, catches ranged between 1266 and 4266 sharks per year, but have since declined. EEC agreements with Norway were limited their catches to 400/yr of livers since 1978. This corresponds to 2 400t/yr whole weight, taking livers as 1/6 of whole weight. Socio-economic constraints which include limited markets and an ageing fleet coupled with erratic distribution of the sharks, are the reasons for the decline of this fishery and this fishery for basking sharks has not taken even the TAC in EEC waters. The oil from the livers is sold for extraction of squalene, a hydrocarbon used in cosmetics and aviation but richer sources have since been found in deep-sea sharks of the genus *Centrophorus* and the market for basking sharks is shrinking. In general, the dynamics of Norwegian elasmobranch fisheries seem to be strongly influenced by economic and social factors (Myklevoll 1989a, 1989b, 1989c). Many of these fisheries in Norway have declined or collapsed for reasons independent of the resource size.

Much about the Norwegian fishery for piked dogfish *Squalus acanthias* in the northeast Atlantic is summarized by Holden (1977) and Myklevoll (1989d). This fishery dates from 1931. Subsequent expansion of the markets led to Norwegian catches of 8767t by 1937 peaking at almost 34 000t in 1963. Since then catches have slowly fallen to less than 6 000t in the 80's. During 1950-1970, Norwegian longliners fished mainly in their coastal waters during winter and in Scottish waters during summer and autumn. The fishery exported most of the catch which was used in fish and chips shops in England. Until the early 70's, this fishery constrained the expansion of the British fishery, due to the larger sizes, better appearance and lower price. In recent years, large numbers of piked dogfish migrated into unusually northern parts of Norway enabling a fishery. This might account for the increase in catches during 1989-91.

During the first half of this century, Norway had a fishery for greenland sharks, *(Somniosus microcephalus)*, both as a specialized activity and in combination with sealing. Judging from the data reported by Myklevoll (1989c) this fishery peaked in 1917 when 17 049 hectolitres of livers were landed. Probably because of falling market prices, the fishery ceased in 1960. Skates and rays have never been exploited a targeted fishery in Norway and all catches are incidental to piked dogfish, ling, halibut and trawl fisheries (Myklevoll 1989e). Species of no commercial value and small specimens are commonly discarded.

Despite developing several specific shark fisheries, Norwegian interest in elasmobranch research have been relatively poor. Of the three most important shark fisheries of Norway (piked dogfish, porbeagle and basking sharks), only the piked dogfish has been studied in any depth in a research programme from 1958 to 1980. This produced the first known assessment of an elasmobranch fishery (Aasen 1964). Aasen estimated a maximum equilibrium yield of 50 000t/yr for what he considered a single stock of piked dogfish for Northern and Western Europe. By 1961, this yield was already surpassed. Porbeagles were briefly studied while the fishery was expanding and this produced one of the first attempts to estimate growth in sharks from vertebral rings (Aasen 1963). There has been only limited research done on basking sharks.

Norwegian vessels fish orange roughy off Australia and New Zealand, but no details about these activities could be found. The use the, probably large, by catches of deep sea sharks from this fishery is unknown (see Section 2.3.4).

2.2.2.2 Former USSR

The elasmobranch fisheries of the ex-USSR were important. Former USSR fisheries for elasmobranchs were not recorded separately from the rest of their fish catches in FAO yearbooks before 1964. Since records began, catches have soared, reaching 59 000t in 1975, declining equally precipitously to about 20 000t in 1977. Since then, catch levels have varied between 10 000-20 000t/yr (Figure 2.2). With the breakup of the Soviet Union, catches plummeted in 1990-1991. Elasmobranchs contributed 0.11% of the total catches for 1987-1991, the lowest among major elasmobranch fishing countries. The contribution to world elasmobranch fisheries by this country was 1.75% in the same period. As for most former USSR fisheries, the elasmobranch catches came from catches of its enormous global fishing activities and a great variety of species are reported under two main headings: rays and various elasmobranchs. The changing characteristics of former USSR fisheries, which largely depended on agreements with various nations, makes their analysis difficult.

Data from FAO (Figure 2.10) show that from 1978 to 1991, rays accounted for 66% (8761t/yr) of the total former USSR elasmobranch catches. Various elasmobranchs represented 31% (4109t/yr) of the catch. Catches of *Squalus acanthias* accounted for the remaining 3% (327t/yr) of the total. Most of elasmobranch catches of the former USSR probably were taken by large trawlers which is shown by their large catches of batoids. Rays were taken mainly in FAO areas 21 (37%), 47 (26%), 27 (15%) and 37 (10%) with the remaining (12%) taken in areas 34, 41, 51 and 71. Catches of various elasmobranchs came chiefly from areas 37 (37%), 47 (31%) and 34 (25%), with the rest (7%) taken in areas 27, 51, 71 and 81. Catches of these two groups in Area 37 consist of thornback ray *Raja clavata* and piked dogfish *Squalus acanthias* fisheries in the Black Sea.

Ivanov and Beverton (1985) indicate that specialized fisheries for these two species are carried out by Crimean and Caucasian fishermen in the Black Sea. Thornback rays are fished with baited longline and caught in bottom gillnets set for piked dogfish. Piked dogfish are also taken by trawl off the northwester coast and by bottom longlines and fixed nets along the coasts of Crimea and Caucasia. After the continuous decline of elasmobranch catches by former USSR fisheries until 1982, catches (mainly of batoid fishes) slowly increased until political events practically shut down all fisheries.

2.2.2.3 United Kingdom

The United Kingdom has one of the most stable elasmobranch fisheries in the world. There has been a steady decrease from 30 000t/yr in the early post-war years to the current level of about 22 000t/yr (Figure 2.2). During 1978-1991, catches varied between 20 000t and 25 000t and are correlated to changes in the catches of piked dogfish *Squalus acanthias* which averaged 63% (13 820t/yr) of total elasmobranch catches (Figure 2.11). Almost 47% percent of piked dogfish catches during this period were caught in England and Wales, with an equal amount caught in Scottish waters. The remaining 6% came from Northern Ireland. Catches of rays averaged 36% (7877t/yr) of all elasmobranchs and have remained fairly constant with a slight tendency to increase. Approximately 49% of ray catches are taken in Scotland and the same amount in England-Wales, while Northern Ireland contributes about 2%. Less than 1% of the total elasmobranch catch of the UK is made up of Scyliorhinids, Squaloids and unspecified elasmobranchs. As a group, chondrichthyans are relatively important to UK fisheries comprising 2.63% of the total catches during 1987-1991.

Holden (1977) summarizes the information for the piked dogfish (*Squalus acanthias*) fishery, which has been fished by England since the beginning of the century but catches did not exceed 2850t until 1931. Scottish catches appeared in records in 1954 and combined catches in UK remained between 6000-10 000t/yr during the 60's and peaked at 19 400t in 1978. During 1950-1970 the amount of spinydogfish caught was dictated by local market demand, and was taken as by catch by trawlers targeting cod, haddock and hake.

According to Kunzlik (1988), fisheries for basking sharks (*Cetorhinus maximus*) existed in the UK during the 40's mainly on the west coast of Scotland. Most were short lived because of marketing difficulties (Maxwell 1952). Basking sharks were hunted mostly during the summer with hand or whaling harpoons from vessels adapted from other fisheries but catches never surpassed 300 sharks per year (approximately 600t/yr). As for Norwegian and other basking shark fisheries in the world, they mainly took livers and present catches are minimal. Since 1983, only one boat fishes, opportunistically, for basking sharks in Scotland.

Porbeagle sharks have been sporadically landed in small quantities (less than 30t/yr), mainly on an incidental basis. The exception was in 1987-1988 when porbeagles were unusually abundant for a couple of months in the Shetland Islands and 35-45t were taken in four months (Gauld 1989).

Although UK catches of skates and rays are larger in the North Sea, most of the available information comes from the Irish Sea. British fisheries for skates and rays in the Irish Sea consist mainly of *Raja montagui, R. clavata, R. brachyura* and *R. naevus* (Holden 1977), in respective order of importance. Fishing pressure has apparently caused a decline in some local stocks. Brander (1977-91) believes that skates and rays of the Irish sea are in need of immediate management measures to allow stock to recovery and attributes the disappearance of *Raja batis*

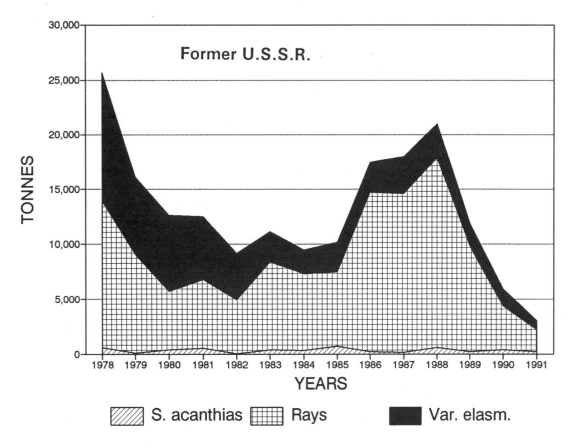

Figure 2.10. Elasmobranch catches of USSR, by species groups, during 1977-1991 (Data from FAO).

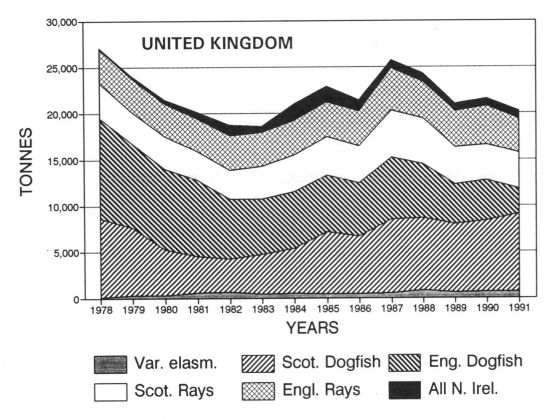

Figure 2.11. Elasmobranch catches of U.K., by species groups, during 1977-1991 (Data from FAO).

from the Irish sea to excessive commercial fishing. According to data summarized in Ryland and Ajayi (1984), stocks of rays in the Bristol Channel, which used to provide 27% of the UK ray catch, were halved during 1964-1974. For the North Sea, Vinther and Sparholt (1988) estimate the biomass of *R. radiata*, and all other rays during the mid 80's, as 160 000-252 000t and 294 000-464 000t respectively. Data presented by these authors suggest declines in the abundance of *R. batis*, *R. clavata*, *R. naevus* and increases in abundance of *R. radiata*. A later biomass estimate of *R. radiata* is 100 000t (Sparholt and Vinther 1991).

Research on elasmobranchs is comparatively active in Britain; however, management seems to be neglected. A fair amount of research was done on piked dogfish (Holden 1968, Holden and Meadows 1962, 1964) but despite the general guidelines proposed by Holden based on his assessment of the fishery, no regulation measures were taken. Also, despite the availability of a reasonable number of basic studies on rays, no management specifically directed to these fishes appears to exist. This might be due, at least partially, to the complications of setting management regulations for multispecific fisheries, especially bottom-trawl fisheries.

2.2.2.4 Ireland

Elasmobranch fisheries of the Irish Republic have been of minor importance until recently, when catches exceeded 10 000t/yr (Figure 2.2). In the period 1987-1991 they contributed 1.03% to the world catch. Despite this small amount, elasmobranch catches are relatively important for Ireland, representing 3.03% of the total landings. This is rather high compared to other major elasmobranch-fishing countries (Table 2.2).

Rays have been long exploited in Ireland in small quantities. Piked dogfish *Squalus acanthias* is the other main elasmobranch resource and has gained much attention since the beginning of the 80's. Since 1983, piked dogfish catches have comprised the major proportion of the total elasmobranch catch (Figure 2.12). During 1978-1991 rays and dogfish were equally represented with catches of 3048t/yr and 3067t/yr respectively. While the catches of rays have remained practically constant since 1978, those for dogfish increased tremendously in less than five years, suffered a small fall in 1986 and recovered and fell again in 3 years. Recent statistics suggest a relative stability has been achieved in this fishery. Fahy (1989a,b, 1991) and Fahy and Gleeson (1990) cover most of what is known about recent elasmobranch fisheries of Ireland and most of the following is taken from them.

Recordings of rays landings of goes back to 1903. No more than 600t/yr was recorded before 1940 when catches began to rise partially due to increased consumption in Ireland, up to the late 70's when they sharply increased, reaching 3 000t in 1985. Rays have traditionally been taken in greatest quantities (around 50% of the total) from the east coast. Since 1975, about 25% has been taken from the north coast and the rest from the south and west coasts. Most of the landings are not sorted by species but are defined by a casual process by similarities in size and appearance. At least 18 trawling vessels catch rays from eastern Irish ports. Thirteen otter trawlers and four beam trawlers operate from the southeast, but more vessels are believed to participate in the fishery. Although most of these vessels catch rays incidentally to prawns and other bottom fish, a small ray fishery appears to occur on a seasonal basis. At least nine species of rays are found in the catches but sampling of the commercial landings indicates that *Raja brachyura*, *R. clavata*, *R. naevus* and *R. montagui* are the most common in order of importance. *R. microocellata*, *R. batis*, *R. fullonica*, *R. undulata* and *R. alba* are sporadically caught. The catch consists mostly of small (less than 60 cm TL) and medium sized rays (between 60-70 cm TL) which account for 60-80% of the weight. Most species are totally recruited to the fishery after 2 years of age but *R. naevus* enters at age 3. At least 50% of the catches of *R. clavata* and *R. brachyura* in the east coast are made of 0-2 age class fish. Total mortality estimates for the most important species range from 0.54-0.74 and although the populations are heavily exploited, particularly in the southeast fishery, they continue to produce good yields.

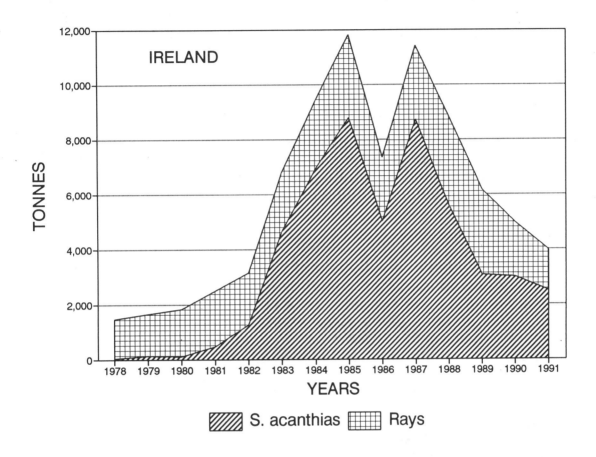

Figure 2.12. Elasmobranch catches of Ireland, by species groups, during 1977-1991 (Data from FAO).

Fisheries for dogfish occur around Ireland country but are more concentrated on the west coast. Catches were high in the north (Co. Donegal) during 1982-1985 but landings in the south (Co. Kerry) increased during 1986-1987 as a result of effort being shifted to the south due to decreasing catches in the north. Dogfish were considered a nuisance but now a fishery is specifically directed at them. On the west coast, otter trawlers fish mainly male dogfish in waters sometimes exceeding 100 m while monofilament gillnets of 6.4cm mesh size are used in shallow waters where they catch high proportions of pregnant females. Piked dogfish in the west of Ireland are fully recruited to the fishery at around 17 years of age and total mortality coefficients have been estimated at 0.24 for females and 0.30 and for males. Fahy and Gleeson (1990) report that monthly CPUE of gillnetters in Carrigaholt plummeted by 80-90% over a two-year period. Available information is insufficient to definitively conclude about the causes of stocks depletion but it seems that they are close to being overfished. Total female spawning biomass for Carrigaholt was estimated at 5700t by Fahy and Gleeson. Most of the catches are destined for export but there is no apparent reason for the boom in this fishery.

A fishery for basking sharks began in 1947 at Keem Bay on the west coast of Ireland (Kunzlik 1988). Initially harpoons and nets were used but by 1951 only encircling nets or entangling nets, set perpendicular to the shore and made of sisal with mesh sizes of 33cm, were used. Initially, the liver was only taken but in later years fins and meat were also used. In 1973 harpoons were reintroduced to this fishery and another harpoon fishery started in the south east coast of Ireland. The west coast fishery peaked (around 1500 sharks annually) during the early 50's and declined after 1955, probably as a response to the shrinking market for livers. Catches remained below 100 sharks/yr during most of the period 1963-1973 and increased to almost 400 sharks in 1975 when the last records are available. Some trials to develop a commercial blue shark fishery with longlines off the south coast of Ireland were done in 1990 (Crummey et al. 1991). Whether a fishery will develop is, as yet, unknown.

2.2.2.5 France

French elasmobranch fisheries are another relatively stable fishery. Two periods of more or less sustained catches exist. From 1948 to 1960, catches oscillated around 15 000t/yr then in 1961 jumped to a higher more variable level around 35 000t/yr (Figure 2.2). During 1987-1991, elasmobranchs represented 3.78% of the total fishery production of France, the highest among European countries and rather high globally. French catches are 4.79% of world elasmobranch production.

Between 1978 and 1991, French catches of skates and "various dogfishes" were stable. Piked dogfish, "various elasmobranchs" and porbeagles showed a slight declining trend (Figure 2.13). During this period, skates averaged 42% (14 499t/yr) of the total elasmobranch catches while piked dogfish, various dogfishes, various elasmobranchs and porbeagles averaged 32% (10 806t/yr), 18% (6139t/yr), 6% (2103t/yr) and 2% (531t/yr), respectively. Piked dogfish and skates are caught by French vessels mainly in the Northeast Atlantic but small catches of skates

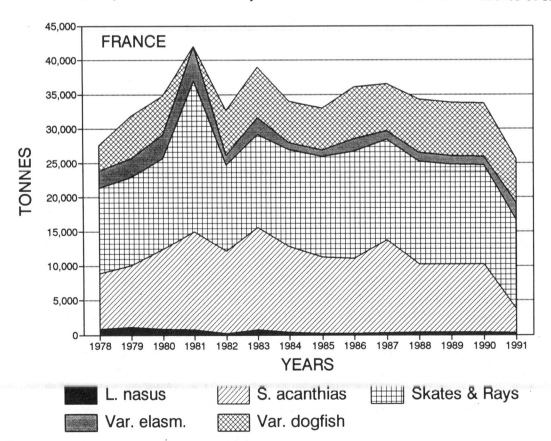

Figure 2.13. Elasmobranch catches of France, by species groups, during 1977-1991 (Data from FAO).

are also taken in the northwest Atlantic and the Mediterranean Sea. According to Gauld (1989), a small flotilla of French vessels based in Britain specifically target porbeagles with longlines in the Bay of Biscay and in Irish waters taking about 75% of the total French porbeagle catch. The remainder is landed as by catch of trawl and seine fisheries.

Tetard (1989a, 1989b) summarizes information about shark and batoid fisheries for France and separates catch statistics into species or species groups. The following is from his account. The catch of batoids of France consists of at least 8 species of skates and rays. Separation of ray species is possible as each species attains a different price. *Raja naevus* and *R. clavata* are the most important accounting for about 25% and 17% respectively of batoid landings during 1978-1987. *Raja montagui* and a group formed by *R. batis* and *R. oxyrinchus* comprise 4% and 3% of the catch respectively. *Dasyatis pastinaca, Myliobatis aquila* and *Raja fullonica* are of minor importance and compose only 1% of the catches. Unidentified rays comprise the remaining 50%. Most of the French catches of rays are taken in waters around the Celtic Sea and the English Channel and to some extent in the Irish sea and the North of the Bay of Biscay. Rays are mostly caught by bottom trawling. *Raja clavata* is actively sought for its highly desired meat. Tetard highlights the almost complete disappearance of *R. alba* from the catches and the apparently declining catches of *R. clavata*. though yields of *R. naevus* seem to be increasing. He also notes that an incited study indicates that the yield per recruit of *R. naevus* is at an optimal value. Judging from Tetards, it appears that no management regulations exist for any of these species in French waters.

Shark landings are chiefly composed of piked dogfish and catsharks. The latter are mainly *Scyliorhinus canicula* with a minor amount of *S. stelaris*. Catshark catches occur as by catch in trawler and longline fisheries and comprise about 32% of the shark catch. The piked dogfish fishery is one of the few directed fisheries for sharks in France accounting for almost 57% of all shark landings. During 1987, approximately 27 longliners 8-25m long (three of them automatic longliners) were targeting piked dogfish. Although, about 80% of the landings came from bottom trawlers. The main fishing grounds for piked dogfish are the Celtic Sea and, formerly, Northern Irish waters, and the North Sea. Tope, (*Galeorhinus galeus*), ranks third in importance among shark catches, with about 6% of the total, but catches are declining. The fishery for porbeagles is also a directed fishery representing about 3% of the shark catch. Some shortfin mako sharks are caught incidentally in the longlines of this fishery. About 75% of the landings come from longliners and the rest from trawlers. The main fishing grounds are offshore waters, from Spain to Ireland in winter, and closer in shore and around the Channel Islands in spring. Smoothhounds, *Mustelus mustelus* and *M. asterias* comprise about 1% of the shark catch. Some minor quantities of blue shark and angel shark, *(Squatina squatina)*, are landed incidentally by longline and trawl fisheries respectively.

France is both the major producer and importer of shark in Europe. High exports of mainly porbeagle and tope shark to Italy results in a deficit of supply and imports have increased since 1982 (9000t in 1986). However, some problems related to mercury content of shark meat seem to limited French exports to Italy, and consequently the effort directed towards porbeagle sharks. The home market is also increasing. There is strong domestic demand for *Lamna nasus*, *Squalus acanthias* and *Galeorhinus galeus* as "saumonette" in schools and restaurants. The domestic demand for *Squalus acanthias* is not met by French landings and considerable quantities are imported from the United Kingdom.

2.2.2.6 Spain

Spanish elasmobranch catches were steady during 1947-1971 when yields varied from 10 000-15 000t/yr. This was followed by a collapse in the early 70's and a subsequent recovery in the 80's to 15 000-20 000t/yr (Figure 2.2). Elasmobranchs comprise 1.3% of the total fishery production of Spain and contribute 1.2% of the world catch (Table 2.2).

Disaggregated data for the years 1978-1991 indicate that the major source of recent increased catches comes from skate fisheries which have grown consistently since 1980 (Figure 2.14). The bulk of skates comes from the Northwest Atlantic (an average, 80% of skate catches for the period) and the rest from the northeastern Atlantic. No information on the species composition is available. Catches of unspecified sharks have also increased in a similar way but these are taken mainly in the Northeast Atlantic. These include shortfin makos (*Isurus oxyrinchus*), porbeagles (*Lamna nasus*), small-spotted catshark (*Scyliorhinus canicula*) and some squaloids. Various species of rays are fished in small quantities mainly in the Mediterranean Sea along with unspecified elasmobranchs which are also caught in the central eastern Atlantic (FAO Area 37). Skates comprise 63% (7125t/yr) and unspecified sharks 21% (2259t/yr) of elasmobranch catches, the contribution of "various elasmobranchs" was 11% (1168t/yr).

All elasmobranch landings in Spain come from incidental catches of trawl or longline fisheries (R. Muñoz-Chápuli, pers. comm., 1992). Muñoz-Chápuli (1985a) reports on the landings of Spanish commercial bottom trawlers operating in depths up to 500m. *Scyliorhinus canicula* dominate landings from the mouth of the Mediterranean, southern Spain and northwest Africa. *Centrophorus granulosus* and *Squalus blainvillei* are also landed from these areas. In the entrance of the Mediterranean, *Galeus melastromus* is also important while another 11 species are caught in smaller amounts in both regions (Table 2.6).

Table 2.6. Shark species reported in Spanish commercial fisheries (adapted from Munoz-Chapuli 1985 a,b).

Demersal	Pelagic
Hexanchus griseus	Lamna nasus
Heptranchias perlo	Isurus oxyrinchus
Squalus acanthias	I. paucus
S. blainvillei	Alopias vulpinus
Centrophorus granulosus	A. superciliosus
C. lusitanicus	Carcharhinus brevipinna
Deania calcea	C. falciformis
Dalatias licha	C. longimanus
Squatina squatina	C. obscurus
S. aculeata	C. plumbeus
Galeus melastomus	C. signatus
Mustelus mustelus	Prionace glauca
M. asterias	Galeorhinus galeus
	Sphyrna zygaena
	S. lewini

Muñoz-Chápuli (1985b) reports that landings from longline vessels fishing from the Azores to the Cape Verde Islands, are dominated by *Prionace glauca, Isurus oxyrinchus* and *Sphyrna zygaena* while another 13 other species are of minor importance (Table 2.6). Both reports likely reflect the abundance of the species in such areas and the species retained on board. Spanish swordfish longliners caught 304t of shortfin makos and 20t of porbeagles from the north and central east Atlantic during 1984 (Mejuto 1985). Makos were more abundant during September-December and catches were mainly composed of sharks 100-240 cm fork length. Males were more than twice as frequent in the catch as females. Porbeagle catches were more abundant in March, September and October. Individuals were mostly 150-225cm fork length.

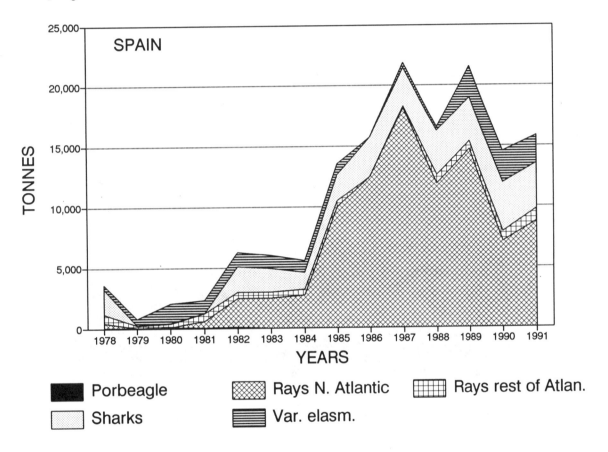

Figure 2.14. Elasmobranch catches of Spain, by species groups, during 1977-1991 (Data from FAO).

2.2.2.7 Italy

The level of historical imports of sharks from Norway (porbeagles), France (porbeagles and tope) and Argentina (smooth-hounds), show elasmobranchs are well appreciated in Italy. Nonetheless, sharks and rays have long been of minor importance in Italian fisheries. Catches did not exceed 6000t/yr until the mid 80's when more than 10 000t/yr were taken (Figure 2.2). Currently, elasmobranchs represent only 1.89% of the total catches in Italy and the Italian catch of sharks and rays comprises only 1.51% of the world elasmobranch catch (Table 2.2).

During 1978-1991, smooth-hounds, (*Mustelus* spp.), averaged 52% (4463t/yr) of elasmobranch catches and rays, 38% (3340t/yr). "Various elasmobranchs" contributed 10% (860t/yr). Catches of all groups grew during the expansion of the fishery which peaked in 1985 (Figure 2.15). Smooth-hounds were all taken from Mediterranean waters along with 91% of the ray catch. The rest were caught in FAO Areas 34, 47, 48, 51 and 21. Catches of "various elasmobranchs" were taken in FAO Area 34 (70%) and Areas 47 (7%), 51 (16%) and 41 (7%). Small catches of blue sharks, *(Prionace glauca)*, are landed as a bycatch of the drift longline swordfish and albacore fisheries of the Gulf of Taranto, where averages of 14.5t/yr and 4t/yr respectively were landed during 1978-1981 (De Metrio et al. 1984). During this period, an average of 12 boats fished for swordfish from April to August using 700 to 1000 (Mustad no. 1) hooks per boat. On average, 44 boats fished for albacore during August to December using 2000 3cm hooks per boat. Due to the different hook size, and probably seasonal cycles of the species, the swordfish boats caught blue sharks of 25kg average weight whereas blue sharks from the albacore boats averaged 3kg. De Metrio el al. report that the meat of *Prionace glauca* is fraudulently sold in Italy as *Mustelus*. It is therefore likely that the blue shark catch is probably reported under *Mustelus* spp. in official statistics

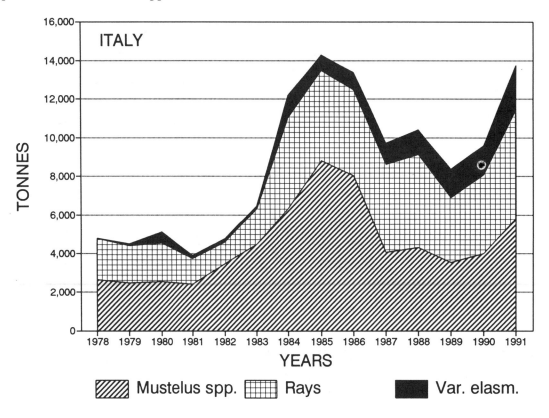

Figure 2.15. Elasmobranch catches of Italy, by species groups, during 1977-1991 (Data from FAO).

2.2.3 Africa and Indian subcontinent

Information about elasmobranch fisheries in this region is scarce. Most of the major elasmobranch-fishing countries give little detail of the catch composition and reports are limited and difficult to obtain.

2.2.3.1 Nigeria

Nigeria is the only African country with a major elasmobranch fisheries. FAO statistics for Nigeria are poor and have only appeared regularly since 1970. They show a fairly unstable fishery with an overall trend of decreasing catches from more than 30 000t/yr in the early 70's to less than 10 000t since 1986 (Figure 2.2). Without background information it is difficult to interpret these figures. Despite the fall in yields, elasmobranchs continue to be a relatively important resource for Nigeria contributing 2.92% of the total fishery production during 1987-1991. The catch of sharks and rays of Nigeria contributes 1.91% of the world total. FAO data from 1977-1991 show that most of the catches are not recorded by species. A group of "various elasmobranchs" accounts for 89% (15 827t/yr) of the catches while Squalidae and a group of skates and rays accounts for less than 1% (7.6t/yr) and about 10% (1703t/yr) respectively (Figure 2.16).

2.2.3.2 Pakistan

Elasmobranch fisheries of Pakistan were of prime importance on a global scale until recently when production plummeted. Elasmobranch landings grew almost exponentially from the late 40's to a peak of about 75 000t in 1973, dropped about 50% during the following three years and then recovering to peak levels for another 6 years. Catches collapsed in 1983 but have recovered during the last 10 years to the present levels of about 45 000t (Figure 2.2). Given the lack of information on Pakistani fisheries it is difficult to determine the reasons for these changes in catches. The relative importance of elasmobranchs in Pakistan is among the highest in the world, 7.42% of the total national catches during 1987-1991. This level must have been at least double during the bonanza of the late 70's. Pakistan landings comprise 4.99% of the world elasmobranch production (Table 2.2).

Batoids and grey sharks (Carcharhinidae) constitute most of the catches, averaging 54% (24 380t/yr) and 45% (20 200t/yr) of the elasmobranch production respectively during 1977-1991. Since 1987, catches of sawfishes (Pristidae) and guitarfishes (Rhinobatidae) have been reported separately, but they account for <1% and 1% of the elasmobranch catches respectively (Figure 2.17). While grey shark catches declined steadily during the late 70's and early 80's batoid catches dropped abruptly by 43 000t in one year (1983) causing the overall collapse. Grey sharks have since been the major species in the elasmobranch catches.

Detailed information about Pakistani elasmobranch fisheries is poor and a report from the Indo-Pacific Tuna Development and Management Programme (IPTP 1991) is almost the only source of information. According to this document, Karachi is the only landing site for the mechanized gillnet fleet in Sind province. Sharks are caught mainly by pelagic gillnet boats fishing as far as Somalia, the Yemen and Oman although small quantities are also landed by bottom gillnetters working in coastal areas of Pakistan. There were 394 mechanized gillnetters in Pakistan in 1989, 185 in Sind province and 209 in Baluchistan. The vessels based in Karachi range in length from 20 to 25m and 5 to 7m in breath and use diesel engines of 88-135 HP. These fisheries are important socio-economically employing considerable numbers of fishermen. Small boats carry 15-17 crew on trips of about 10 days; larger boats carry up to 25 fishermen for 20-30 days and occasionally 60 days. Catches are usually salt dried on larger vessels and kept on ice in the smaller ones. Gillnets are hand-woven out of multifilament polyamide twine and are 80 meshes deep and 2.5-9km long (average of 5.2km). Mesh sizes are 10-16cm and mainlines of 14-16mm diameter. Sharks are categorized into 8 types depending on size and species. Effort in this fishery increased from 23 000 fishing days in 1988 to 28 000 in 1989 then fell to 26 000 days in 1990. About 93% of the shark catch comes from pelagic driftnet vessels. The production of sharks of this driftnet fleet was about 3860t/yr during 1988-1990. Shark production during this period was correlated with distance to fishing grounds. The greatest catches came from Somalian waters, the most distant fishery. Shark yields decreased by 44% from 1989 to 1990 while other catches fell 32% during the same period. Some efforts to introduce longline fishing for sharks, rays and other species in Pakistan are described by Prado and Drew (1991). Apparently gillnets are more favoured in Pakistan because of their higher catch rates of valuable species.

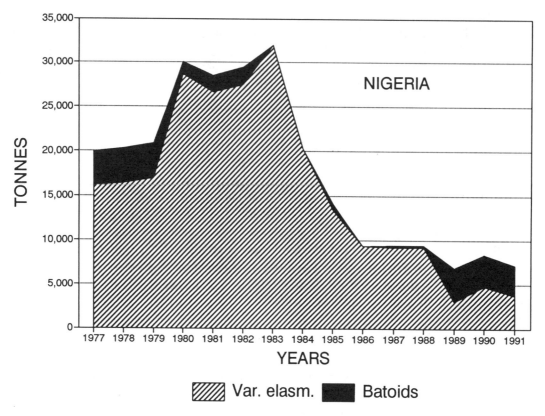

Figure 2.16. Elasmobranch catches of Nigeria, by species groups, during 1977-1991 (Data from FAO).

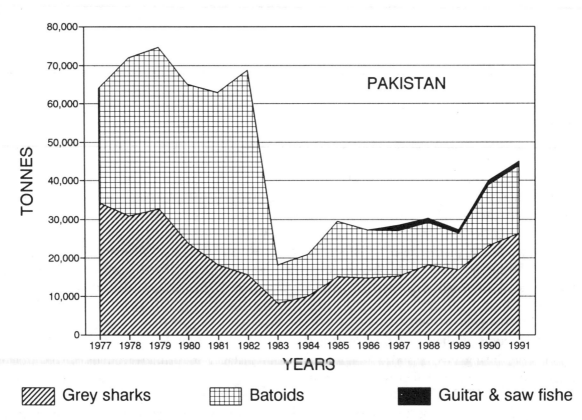

Figure 2.17. Elasmobranch catches of Pakistan, by species groups, during 1977-1991 (Data from FAO).

2.2.3.3 India

There have traditionally been important fisheries for elasmobranchs in India with a relatively steady growth up to the mid 70's, followed by a period of stability during most of the 80's, then a tremendous increase in catches in 1987 resulting in India becoming one of the top three elasmobranch producers in the last ten years (Figure 2.2). Indian production of sharks and rays represents 8.78% of the world elasmobranch catches! Still, because of large inland yields, elasmobranchs comprise only 1.72% of total national catches in 1987-1991. Catches results are not given by species or families in the statistics and the composition of catches is only known by FAO areas. Approximately equal amounts (about 26 000t/yr) were obtained from both FAO areas for the period 1977-1991. Catches from the west coast were slightly larger than those of the east coast during 1977-1991 (Figure 2.18). There is a relatively large number of articles on elasmobranchs' exploitation and utilization in India, especially for the 80's.

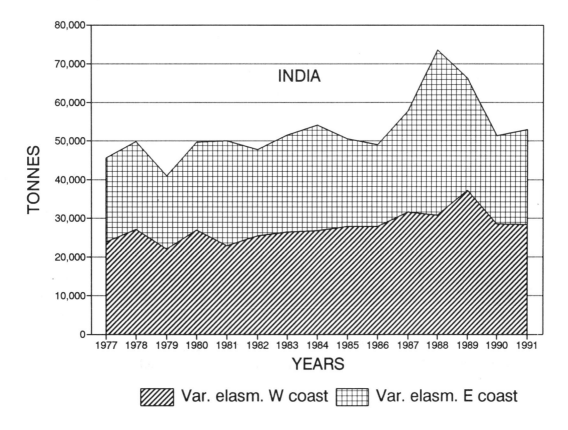

Figure 2.18. Elasmobranch catches of India, by species groups, during 1977-1991 (Data from FAO).

During 1983-1985 sharks comprised 55% of the elasmobranch catch of the country (Appukuttan and Nair, 1988). The main fishing areas in order of importance were Gujarat, Maharashtra, Kerala andhra Pradesh, Karnataka and Tamil Nadu and important fishing grounds for sharks are reported for Ashikode, Kerala Province (Anon. 1983). Sharks catches are incidental to other fisheries in India (Appukuttan and Nair 1988) and are mainly taken with longlines, which vary in design by region, and are also as by catch of trawlers using disco nets off Ratnagiri (Maharashtra), with bottom set gillnets in Porto Novo (Tamil Nadu) and by shrimp trawlers of Kerala (Devaraj and Smita 1988; Shantha et al. 1988; Rama Rao et al. 1989; Kulkorni and Sharangdher 1990). Rays are caught with bottom set gillnets in Gujarat, northwest India and Cudalore and are abundant on the outer shelf and slope off Kerala and Karnatakta (Devadoss 1978; Kunjipalu and Kuttappan 1978; Sudarsan et al. 1988). Devadoss (1984) indicates that batoids comprise 10% of by catches in Calicut; 90% of the by catch comes from trawlers, 8% from gillnets and 2% from hook and lines. Both sharks and rays are abundant in Lakshakweep and form important by catches in trawl fisheries in Krishnapatnam (Swaminath et al. 1985; James 1988).

Dahlgren (1992) notes that directed fisheries for sharks are developing on a seasonal basis on the east coast of India. About 500 vessels, both sail-powered and motorized, fish for sharks with bottom or drift longlines of the coasts of Orissa Andhra Pradesh and Tamil Nadu. Bottom longlines are usually set in waters 80-150m deep and occasionally as deep as 500m and bull sharks and tiger sharks. The longlines have up to 400 hooks and the meat is usually salted on board during the trip. In Orissa alone, about 200 boats are engaged in drift longlining on a seasonal basis (December-March). The most common species caught by drift longlines are silky sharks and scalloped hammerhead sharks.

Catch composition data are not readily available but the multispecies nature of these fisheries is evident from the literature. Appukuttan and Nair (1988) report that more than 20 species of sharks (mainly carcharhinids and sphyrnids) are commonly caught. Their data for Pamban and Kilakkarai show that *Rhizoprionodon acutus*, *R. oligolinx*, *Carcharhinus limbatus*, *C. sorrah*, *C. hemiodon*, *Sphyrna lewini* and *Eusphyra blochii* are the most important species. Other species caught are *C. melanopterus* and *Scoliodon laticaudus* (Devadoss 1988). Important batoids are: *Dicerobatis eregoodoo*, *Rhynchobatus djiddensis*, *Rhinobatus granulatus*, *Himantura uarnak*, *H. bleekeri*, *Dasyatis sephen*, *D. jenkinsii*, *Aetobatus narinari*, *A. flagellum*, *Aetomylus nichofii* and *Mobula diabolus* (Devadoss 1978, 1983; Kunjipalu and Kuttappan 1978).

Local assessments of the state of the fisheries for elasmobranchs exist (Santhanakrishnan 1983, Krishnamooorthi et al. 1986, Devadoss et al. 1988, Sudarsan et al. 1988), but no overall studies exist (Appukuttan and Nair 1988). Devadoss (1983) reports that ray resources off Calicut were apparently overfished by 1980 while according to Reuben et al. (1988) shark and ray resources of Northeast India were still underexploited in 1985. Devadoss et al. (1988) did local assessments using Schaefer's model and made suggestions for effort changes for the different areas. The present situation needs careful monitoring. There appears to be a high level of catches of elasmobranchs in India (peak of 73 500t in 1988) and it is unlikely that such large yields are sustainable over a long periods. The collapse of the neighbouring Pakistani elasmobranch fisheries in 1983 could indicate future catch reductions for the Indian elasmobranch fisheries.

2.2.3.4 Sri Lanka

Statistics for the elasmobranch fisheries of Shri Lanka exist since the early 50's. The fishery development has been slow, growing from less than one tonne in 1952 to about 15 000t/yr (Figure 2.2). These fisheries are the smallest among major elasmobranch-fishing countries in the Indian Ocean. Despite this, elasmobranchs are important nationally, contributing 8.76% of the total catches during 1987-1991. This is the highest percentage importance of any elasmobranch fishery in the world. The catch of sharks and rays of Sri Lanka represents 2.42% of the world elasmobranch catch for the period 1987-1991 (Table 2.2).

Information on catch composition is poor for Sri Lankan elasmobranch fisheries. FAO data indicate that catches were commonly grouped in a single "various elasmobranchs" category until 1987.

Since then the category "*Carcharhinus falciformis*" constitutes the major part of the catch (Figure 2.19). But, information from the National Aquatic Resources Agency (NARA) of Sri Lanka (P. Dayaratne, NARA, Colombo, Sri Lanka, pers. comm. 1992) indicates that *C. falciformis* comprises 75% of the shark catches, with *C. longimanus, C. sorrah, Sphyrna lewini, Alopias pelagicus* and *Isurus oxyrinchus,* high among the remaining 25%.

There are few directed fisheries for elasmobranchs in Sri Lanka. Some estimates (P. Dayaratne, pers. comm. op. cit.) indicate that approximately 85% of the elasmobranch caught is by catch from other fisheries which use mainly bottom and drift gillnets. Both the directed and incidental catches of elasmobranchs come from small-scale fisheries. Drifting shark longlines are used in offshore (>40km from shore) EEZ waters in the directed fishery. Bottom set gillnets operate in coastal areas up to 25km from shore (P. Dayaratne pers. comm. op. cit.). Pajot (1980) reports 26.62% the total catch weight from large-mesh small-scale driftnets off Sri Lanka, consists of sharks and rays. There is some detailed information about the pelagic tuna fisheries off Sri Lanka which catch substantial amounts of sharks. Most of the available information is from the IPTP/NARA tuna sampling programme. The following summary is based mostly on the reports of IPTP (1989), Dayaratne and Maldeniya (1988), Dayaratne and de Silva (1990) and Dayaratne (1993a,b).

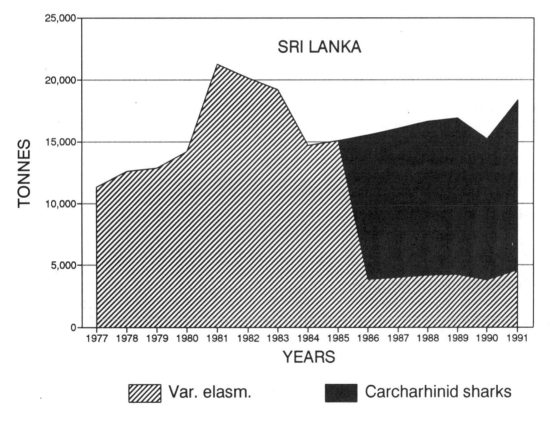

Figure 2.19. Elasmobranch catches of Sri Lanka, by species groups, during 1977-1991 (Data from FAO).

The sampling programme was initiated in Kandakuliya in the northwest, Negombo in the west and Beruwala in the southwest coast of Sri Lanka during 1986 and was extended to two additional locations Matara and Hambantota in the south coast, in 1987. Three types of vessels operate in the pelagic tuna fisheries: small outboard motor boats of about 5m length, diesel motor vessels of about 9m length and 3.5t displacement and the larger 11m long, 11t net tonnage vessels with inboard diesel motors. By far the most numerous are the 3.5 GT vessels numbering about 2000 vessels. They usually carry a crew of four and about 40 panels of net. There are over 1000 of these boats which spend more than 1 day offshore for trip. In contrast, there are only 70 of the 11 GT boats but these usually carry 50-60 panels of net and are capable of making offshore trips of 6-8 days. Gillnets are the most popular gear and they have been used for decades by Sri Lankan fishermen. Each piece of net has 500 x 100 meshes which are of 90-180m, most commonly 140-152m, making a total of 3-4.5km of net per vessel. Overall, the yield and catch rate of sharks in this fishery are variable but both show an increasing trend. Total shark catch grew from 1569t in 1986-1987 to 2155t in 1987-1988 in the northwest, west and southwest coasts. For the west and south coasts, total shark catches increased from 3159t to 4374t, to 8676t during 1989-1991. Overall, shark catch rates increased from about 10 kg/day/boat in 1986 to about 35-40 kg/boat/day in 1988. These increases in shark yields and CPUE reflect trends seen in the fishery which include expansion of fishing to offshore areas, increase in time spent at sea and a change in fishing gear to involve fewer vessels fishing solely with gillnets and more switching to multiple-gear fishing. The percentage importance of sharks in the catch of each gear combination is 15% for driftnets, 28% for vessels using driftnets/longlines/handlines, 40% for driftnets/longlines/troll lines and 45% for driftnet/longline vessels. Elasmobranch catches for each gear type in 1991 were: driftnet 313t; driftnet/longline 3569t; driftnet/longline/handline 513t and driftnet/longline/troll line 1110t. The sharks in the pelagic tuna fishery are dominated by grey sharks (Carcharhinidae) which constitute 85% of the shark catch, hammerhead (3.5%), thresher sharks (1%), mackerel sharks (0.7%) and other sharks and rays comprising the remaining (10.3%). The weight of sharks is estimated visually. There are plans to include three species of sharks (*Carcharhinus falciformis, C. longimanus* and *Prionace glauca*) in the field sampling soon (J. Morón, IPTP, pers. comm. 1993).

In Sri Lanka, at present, there are neither management measures for these fisheries nor are any being considered. So far, there is no evidence of conservation problems or of any species being endangered. Nonetheless, data show that sharks and rays represent an important fishery for Sri Lanka and they should be carefully managed. This summary shows that at least the pelagic fishery is presently in a developing stage.

2.2.4 Asia

2.2.4.1 Japan

Statistics show that Japan catches the world's largest amount of elasmobranchs. Catches have followed a decreasing trend after an initial explosive growth from the late 40's when a record 118 900t were caught (Figure 2.2). Despite this reduction, Japan's elasmobranch production of 37 000t was among the top seven in the world in 1991 contributing 4.98% of the total world catch for the period 1987-1991. This is still high when compared with most other countries. Taniuchi (1990) reports that the relative importance of sharks (which traditionally comprise the majority of elasmobranch catches) dropped from 4.3% of the total fish catches in 1949 to 0.3% in 1985 and that both a decline in the relative value of elasmobranchs and a reduction of the Japanese elasmobranch stocks seem responsible for the decline. At present, elasmobranchs constitute 0.31% of the total Japanese catches, one of the lowest among major elasmobranch-fishing countries (FAO data for 1987-1991). Taniuchi also reports a sharp reduction in catches of *Squalus acanthias* in Japan from more than 50 000t in 1952 to less than 10 000t in 1965 and that this likely represents a reduction of the species' stock-size as catches of other sharks did not follow the same trend. However, stock reduction is not the only factor causing Japan's reduced harvests. As the economy of the country grew during the post-war period, changes in purchasing power will have modified consumer preferences which could also change demand for elasmobranchs. This trend is confirmed by the large amounts of sharks that are discarded by various Japanese fisheries.

Japanese elasmobranch production is chiefly a bycatch of other fisheries. Some exceptions are a trawl fishery for skates and rays in the East China Sea, a salmon shark fishery off northeast Japan in the Oyashio Front (Paust, 1987) and a winter fishery in Hokkaido for *Raja pulchra* (Ishihara 1990). Additionally, small scale coastal gillnet fisheries takes up to 3817t of sharks which accounts for less than 0.01% of the total coastal gillnet catch in Japan (Anonymous 1986). Several trends occur in the data given by Taniuchi (1990) and Ishihara (1990) for the period 1976-1985 (Figure 2.20). Sharks accounted for 83% of the elasmobranch catches of Japan and batoids for 17%; at least 63% of the shark catches were taken as by catch of world-wide tuna longline operations while the remaining 37% came from unspecified sources. Of the average 25 000t/yr of sharks landed by the tuna longline fleet, 58% came from offshore areas, 33% from the high seas and only 9% from coastal waters, presumably the Japanese E.E.Z. The data also show that a shark catch equivalent to approximately 2.8 times the landed shark bycatch of the longline tuna fishery is discarded at sea. Of the approximately 9000t/yr catch of batoids, 50% were caught in the East China Sea, 35% in Hokkaido and 8% in the Sea of Japan. Japan has some of the largest high seas fisheries for tunas and billfishes in the world. These produce substantial bycatches of sharks, only some of which are utilized (See Section 2.3).

Data from FAO for 1977-1991 indicate that sharks are taken mainly in the northwest Pacific (Area 61) where Japanese catches are rapidly declining (Figure 2.21). Approximately 8000t/yr are taken in the rest of the Pacific; this catch has a fairly constant trend and small amounts are also taken in the Indian and Atlantic Oceans. All batoid landings come from the northwest Pacific.

Detailed data on the species composition of the catches are not available from Japanese statistics after 1968. However, Taniuchi (1990) gives data for 1951-1967 and reports piked dogfish *Squalus acanthias* as the main species in the catch up to 1958 followed by blue shark *Prionace glauca* and salmon shark *Lamna ditropis*. The same paper lists 25 shark species captured by tuna longline vessels. Considering the current importance of shark bycatches in longline fisheries to the total shark catch, and data from research cruises reported by Taniuchi (1990), the most important species in the shark catches at present should be, in order, the blue shark *(Prionace glauca)*, the silky shark *(Carcharhinus falciformis)*, the oceanic whitetip shark *(C. longimanus)* and the shortfin mako *(Isurus oxyrinchus)*. However, discarding practices at sea and the nature of the remaining shark catch that does not come from tuna longliners might affect this. In the East China Sea, *Raja boesemani, R. kwangtungensis* and *R. acutispina* are respectively the most important species in the batoid catch (Yamada, 1986).

The meat and the cartilage of Elasmobranchs are used in Japan for traditional dishes, industrial and medicinal uses of liver oil compounds and the skins for making leather. However, Japanese fishermen consider sharks a nuisance as they damage gear and eat hooked tunas and billfishes, and even as competitors for exploitation of valuable fish stocks (Taniuchi 1990). No management measures are known for elasmobranch fisheries in Japan.

2.2.4.2 South Korea

The records of South Korean elasmobranch fisheries are intermittent and limited to FAO statistics. South Korea has taken more than 10 000t/yr of elasmobranchs since at least 1948 and yields show an increasing trend varying around 20 000t/yr since the mid-80's (Figure 2.2). Their recent catch of sharks and rays contributes 2.67% of the total world elasmobranch catch (Table 2.2). Given the large fisheries production of South Korea, elasmobranchs are of minor importance representing only 0.66% of the total catches (1987-1991).

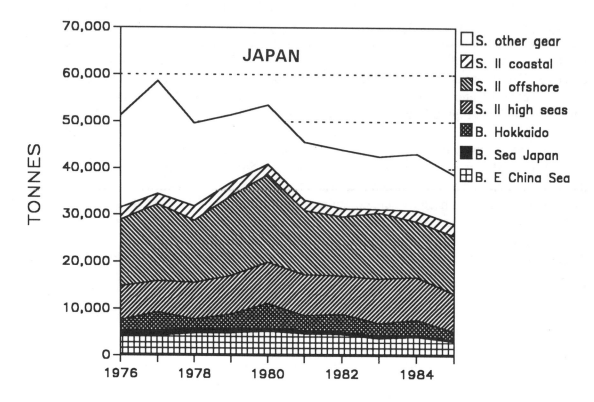

Figure 2.20. Elasmobranch catch in different fisheries of Japan during 1976-1984 (S=sharks, B=batoids, ll=longline) (Data from Taniuchi (1990) and Ishihara (1990)).

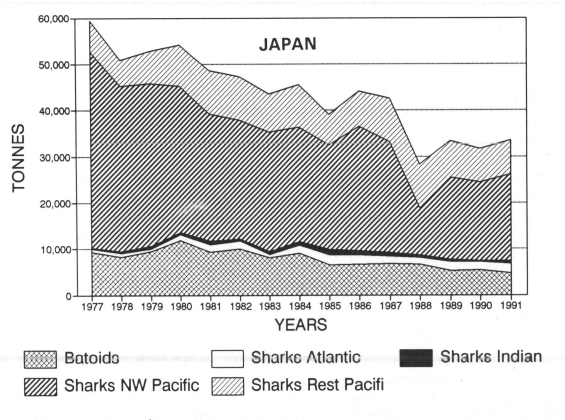

Figure 2.21. Elasmobranch catches of Japan, by species groups and region, during 1977-1991 (Data from FAO).

The elasmobranch fisheries of this country are poorly documented - there are no reports on catch composition by species. FAO data (1977-1991) identified two major categories, batoids and "various elasmobranchs." The latter probably refer to sharks (Figure 2.22). During this period batoids constituted 73% of the elasmobranch catch and were taken chiefly in the Pacific Ocean (94%), with small catches in the Atlantic (4%) and the Indian Oceans (<1%). Other elasmobranchs came mainly from the Pacific Ocean (88%) and in small quantities from the Atlantic (9%) and Indian Oceans (3%). Although batoids represent the major proportion of the elasmobranch catch according to FAO statistics, the data represent only the actual landings and not discards. South Korean markets may, to some extent, influence the discard procedures at sea. The Korean longlining tuna fleet is known to catch and probably discard great numbers of sharks on the high seas of the world (see Section 2.3).

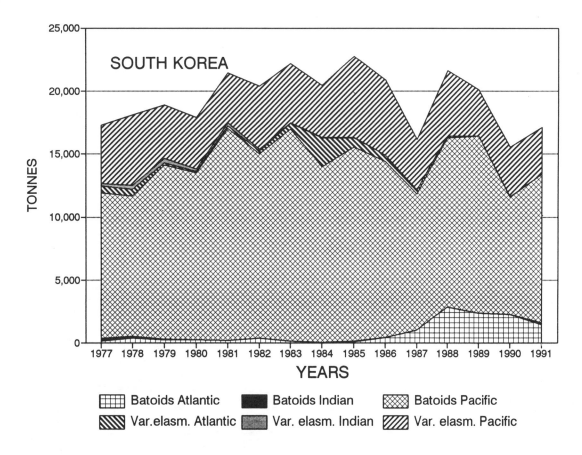

Figure 2.22. Elasmobranch catches of South Korea, by species groups and region, during 1977-1991 (Data from FAO).

2.2.4.3 People's Republic of China

No information on the elasmobranch fisheries of the People's Republic of China exists in FAO statistics. The fisheries agency in China says that no information on elasmobranch fisheries exists. However, China has been exporting increasing quantities of shark fins to Hong Kong during the past few years so that a harvest of sharks must exist, even if as an incidental catch. A rough estimate based on data from the Southeast Asian Fisheries Development Center (SEAFDEC) on shark fins exports to Southeast Asian countries (P. Wongsawang, SEAFDEC, Samutprakan, Thailand, pers. comm. 1992) indicates that China's shark catch apparently grew from less than 100t in 1981 to between 17 000t and 28 000t in 1991, depending on which conversion factor is used (Figure 2.23). These are minimum estimates as an unknown part of the production might not be exported. Actual catches should be much higher. According to Cook (1991), due to the recent relaxation in import and consumer restrictions in China, demand for the traditional shark fin soup has soared, creating extra demand for the product. In addition to the expansion of imports mentioned by Cook, this must be causing increased exploitation of elasmobranchs.

Zhow and Wang (1990) provide some information confirming the existence of fisheries for sharks and rays in the People's Republic of China and give some details. Sharks and rays are caught using driftnets, set gillnets and longlines (there are more than 3.5 million gillnets are used in China) Driftnets range from 30mm to 360mm mesh size but probably those targeting elasmobranchs are near the upper limit of this range. Driftnets target sharks in Xiapu and Jinjiang, Fujian Province. Set gillnets occur in mesh sizes of 30-320mm and are used in shallow waters to target, among many other species, *Triakis scyllium* and *Squalus fernandinus* in Haiyang, Shandong Province. Set longlines of different types are used to catch various elasmobranchs. They vary between 388 and 500m in length. *Prionace glauca* and *Carcharhinus* spp. are targeted with longlines in Hui'an, Fujian Province, "various sharks" are caught in Yangjiang, Guangdong Province and "various rays" in Changdao, Shandong Province. A variation of longlines called rolling lines ares used to catch rays in Haixin, Hebei province, Minhou, Fujian Province and Rudong, Jiangsu Province. These consist of non-baited sharp hooks narrowly spaced on the main line.

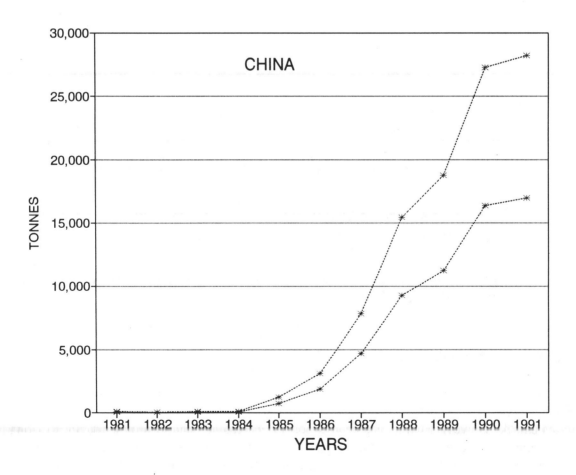

Figure 2.23. Estimated shark catches for the People's Republic of China from fin exports, using 3% and 5% conversion rate (P. Wongsawang, pers. comm.).

2.2.4.4 Taiwan (Prov. of China)

Taiwan (Prov. of China) has one of the world's most important elasmobranch fisheries oriented mainly towards sharks. No comprehensive information on elasmobranch catches before the 70's could be found for Taiwan (Prov. of China) but data from the Fisheries Yearbooks of Taiwan (Prov. of China) Area show that large quantities of elasmobranchs have been harvested since the 1950's (Figure 2.2). Total elasmobranch catches fluctuated around 45 000t/yr during 1979-1988. This was followed by a substantial increase of catches in 1989 and especially 1990 when production reached more than 70 000t as a result of increased catches of large sharks (Figure 2.24). These variations probably represent changes in discard practices of the fleet rather than expanded effort. Elasmobranchs comprised 3.5% of the total catches of Taiwan (Prov. of China) from 1987-1991. Large sharks constitute the majority of the catches, approximately 81% of the total elasmobranch catch during 1978-1990. Small sharks account for approximately for 14%, while rays are of little importance contributing about 5%. Main elasmobranch species in the catch are hammerhead sharks (*Sphyrna lewini*, *S. zygaena*), grey sharks (*Carcharhinus plumbeus*, *C. falciformis*), mako sharks (*Isurus oxyrinchus*), blue sharks (*Prionace glauca*) and thresher sharks (*Alopias superciliosus*, *A. pelagicus*) (C.T. Chen, National Taiwan Ocean University, pers. comm. 1992).

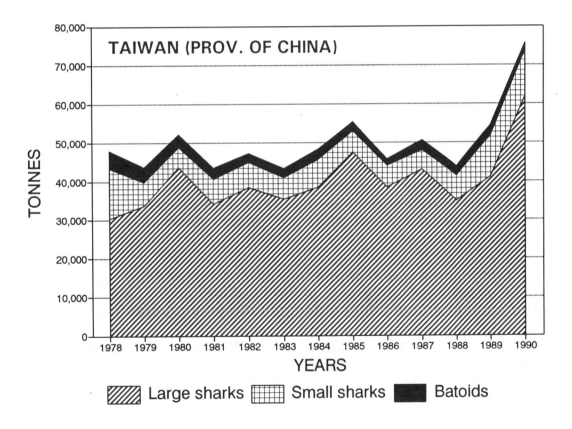

Figure 2.24. Elasmobranch catches of Taiwan (Prov. of China) by species groups, during 1978-1990 (Data from FAO).

Most of the shark catch from Taiwan fisheries are obtained outside their own waters by the various far-seas tuna fleets. During 1988-1990, approximately 85% of the large shark and 70% of the small shark catches came from these operations. In contrast, most of the ray catch (53%) for the same period were caught in Taiwanese waters.

The Taiwanese far-seas fleet is difficult to monitor as it operates in all the oceans of the world and is composed of many sizes and types of vessels (i.e. longliners, driftnetters, purse seiners) (Ho, 1988). Significant shark catches are taken by large-scale driftnetters targeting sharks particularly in Indonesian waters of the Arafura, Banda and Timor Seas.

Taiwan (Prov. of China) prosecuted an important fishery for sharks in Northern and North Western Australia waters from 1972 to 1986 for which Millington (1981) Okera et al. (1981) and Stevens (1990) provide some information. This was mainly composed of driftnetters setting multifilament nylon nets varying between 3 and 12km length, 140-190mm mesh size and 17-30m deep. Vessels size ranged between 160 and 380 GT. Further, Taiwanese pair trawlers fishing for demersal fish took shark bycatches on approximately the same grounds as the driftnetters. The catches of driftnetters were 80% sharks. Of these, *Carcharhinus tilstoni* and *C. sorrah* were the main component (55% of total catches), the remaining were tuna and mackerel. Between 3500 and 14 800t/yr of sharks were taken by these driftnetters during the period 1975-1980. Catches from pair trawlers averaged approximately 2000t/yr of sharks; up to 7000t were taken in 1974. Limits on the number of vessels, and fishing areas and a catch quota of 7000t were imposed on this fishery in 1979 by the Australian Government. The Taiwanese shark driftnet fleet left the fishery in 1987 following the imposition of a maximum gillnet length of 2.5km by the Australians which made the fishery unprofitable (Stevens 1990) but have since continued the fishery in Indonesian waters. At least 7000t/yr of sharks were taken by the Taiwanese fleet in the Australian EEZ before 1987. It is unknown how much they presently catch in Indonesia. If the SEAFDEC figures reported for Taiwanese large-scale gillnet shark catches correspond to the fishery in Indonesian waters, then 19 636t were taken there in 1987. Also, bycatches of sharks in other important large-scale Taiwanese fisheries, for example the tuna longline fishery, the Indian Ocean driftnet fishery and North Pacific squid driftnet fishery, must account for part of the shark catches of this country but are so far unrecorded. These fisheries are further discussed in Section 2.3.

Data from the Fisheries Yearbooks of Taiwan Area, during 1988-1990 show that the main fishing localities for large sharks were Ilan Hsien and Pingtung Hsien. These areas account for 32% (2109t/yr) and 49% (3246t/yr) of the large sharks caught in Taiwanese waters. Keelung Hsien was the main site for catches of small sharks and rays providing 37% (991t/yr) and 73% (875t/yr) of the local catches of each group respectively.

Most of the Taiwanese shark catches are taken by large-scale fisheries, particularly with longliners. According to SEAFDEC data, about 90% of the domestic elasmobranch catch of 9529t (those taken in the South China Sea Area) in 1988 came from large-scale fisheries. For sharks, large-scale longlines and hook and lines accounted for 62% of the catches while gillnets and otter trawls accounted for less than 20% each (Table 2.7). Only 5% of the shark catch came from small-scale gillnet fisheries and less than 1% from traps and longlines. For rays, otter trawls were the most important large-scale gear with 23% of the catch, but gear classified as large-scale "others" took 58%. Gillnets took to 7% of the small-scale catch. The remaining 11% of ray catches was taken using small-scale gillnets and traps.

It is unknown if any stock assessment has been done for the Taiwanese fisheries. Nevertheless, elasmobranch stocks in Taiwan (Prov. of China) are believed to be overexploited and tiger sharks (*Galeocerdo cuvieri*) are considered to be an endangered species (C.T. Chen, pers. comm. op. cit.). Despite this, no management measures exist or are being considered for Taiwan's elasmobranch fisheries.

Table 2.7. Percentage catches of sharks and rays according to fishing gear and zones in Taiwan (Prov. of China) and Malaysia (data from SEAFDEC 1988).

TYPE OF FISHERY AND GEAR	TAIWAN SHARKS	TAIWAN RAYS	PENINSULAR MALAYSIA WEST COAST SHARKS	WEST COAST RAYS	EAST COAST SHARKS	EAST COAST RAYS	INSULAR MALAYSIA SABAH SHARKS	SABAH RAYS	SARAWAK SHARKS	SARAWAK RAYS
LARGE SCALE										
Purse seine	-	-	0	-	0	-	-	-	-	-
Trawl	-	-	63	80	70	93	60	72	-	-
Otter trawl	11	23	-	-	-	-	-	-	30	70
Gill net	17	7	-	-	-	-	-	-	-	-
Hook & line	62	0	-	-	-	-	-	-	-	-
Others	4	58	-	-	-	-	-	-	0	0
SMALL SCALE										
Gill/drift net	5	4	28	4	20	5	15	-	54	11
Hook/long line	0	-	8	16	9	0	25	26	15	17
Trap	0	7	-	0	0	0	0	0	1	2
TOTAL CATCH (mt)	8588	941	1359	6125	1111	2303	910	596	1872	2546

2.2.4.5 Philippines

Philippine's elasmobranch catches were of minor importance before the late 1970's and although variable, expanded until 1986 stabilizing around 17 000t/yr (Figure 2.25). From 1987-1991 they comprised only 0.8% of the total national catches. SEAFDEC data show rays to be slightly more important than sharks in the catches representing an average 53% of the elasmobranch yields during 1977-1991, although both groups showed a growing trend the catches during this period. Philippine catches account for 2.63% of the worldwide elasmobranch catch.

Judging from the 1988 catches (17 879t), small scale fisheries provide the large majority of elasmobranch catches in Philippines (Table 2.8). In Luzon, large scale trawlers accounted for 30% of the local shark catches but only 6% of rays, with purse seiners taking around 3% of both groups' catches. In Visayas, trawls were the main gear in large scale fisheries for rays (23%) but accounted for only 1% of that of sharks. Large scale purse seining took 11% and 8% of the shark and ray catches respectively in that area. Catches from small-scale fisheries for both sharks and rays in Luzon and for sharks in Visayas were mainly taken by hook and line and longlines (38%-76%) but also by gillnets (8%-30%). The reverse was the case for catches of rays in Visayas where gillnet catches were greater than those from hook and line and longline (42% vs. 22%). Small contributions to the catches of both fishes were made in Visayas and Luzon by "other gear" (< 13%). Minor catches of rays were also taken with traps (< 8%). Small scale fisheries took all the elasmobranch catches in Mindanao. The main gear was with gillnets in the case of rays (81%) and hook and line for sharks (57%). Small scale gear, classified as "other", were the second most important method of catching both groups (28% of sharks, 10% of rays). Gill nets took 15% of the small-scale shark catches and traps less than 1%. For rays, hook and line and longlines were the third most important gear in this area taking 7% of the catch. Traps and otter trawls took little.

The composition of batoid and shark catches by area is shown in Figure 2.25 based on SEAFDEC data. Mindanao is the most important area for the catches of both sharks and rays, averaging 3185t/yr (24% of total elasmobranch catches) and 2724t/yr (21%) respectively. Ray catches have generally grown there while shark catches have been variable. Luzon is the second area in importance with 1993t/yr of sharks (15%) and 2312t/yr of batoids (18%). Shark catches in Luzon have decreased from the levels of the late 70's while batoid yields have recently increased after a decline in catches in the early 80's. Production of sharks and rays in Visayas is the lowest in the Philippines with averages of 1108t/yr (8%) and 1856t/yr (14%) respectively; yields of both groups show the same behaviour for batoid catches in Luzon.

Little is known about the species composition of elasmobranch catches in the Philippines. Warfel and Clague (1950) report tiger sharks to be the major catch of shark longlines around the Philippines from exploratory fishing. Other sharks found in the survey include at least six species corresponding to the genus *Carcharhinus*, plus *Sphyrna zygaena*, *Scyliorhinus torazame*, *Hexanchus griseus* and an unidentified nurse shark. The species taken by gillnets were *Pristis cuspidatus* and *Rhynchobatus djiddensis*. Encina (1977) reports on a new dogfish fishery catching *Squalus acanthias* and *Centrophorus* spp. around the Philippines, primarily directed prosecuted for squalene oil extraction.

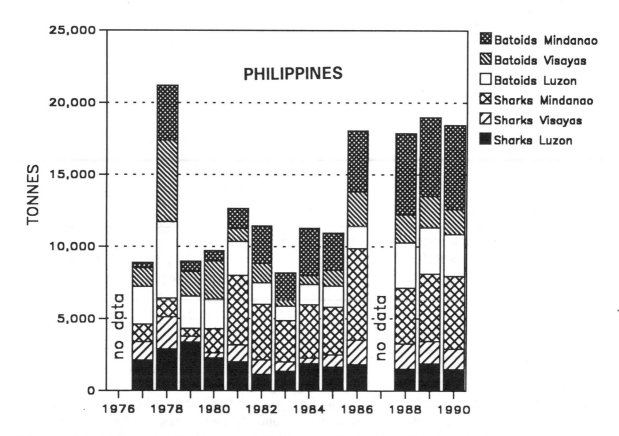

Figure 2.25. Elasmobranch catches of Philippines, by species groups and region, during 1976-1990 (Data from SEAFDEC).

2.2.4.6 Thailand

Now, one of the more modest major elasmobranch fishing countries in Southeast Asia, Thai catches grew considerably in the 1960's but have declined since the early 1970's (Figure 2.2) mainly as a consequence of over-exploitation by trawlers in the Gulf of Thailand (Menasveta et al. 1973, Pope 1979). In later years, there were signs of an apparent recovery but catches have, since 1988, dropped again and the present state of the stocks is uncertain. Sharks and batoids represent a minor resource in Thailand and contributed only 0.43% of the total production during 1987-1991.

Rays, taken as a bycatch by trawlers, dominate the elasmobranch catches. SEAFDEC data show that average catches of rays for the period 1976-1991 accounted for 64% of the elasmobranch production, while sharks were only 36%. Estimates of the Thai Department of Fisheries show that approximately 95% of the shark catch is composed of individuals smaller than 1.5m TL, mainly *Carcharhinus* spp., while the main batoid species in the catch are *Dasyatis* spp. and various eagle rays. (P. Saikliang, D.O.F. pers. comm. 1991).

Thai elasmobranch fisheries are chiefly a large-scale activity. Of a total of 11 438t of elasmobranchs taken in 1988 by Thailand, most of the catches on both coasts of the country came from large-scale trawlers. Otter trawls caught 63% and 82% respectively, of the shark and ray catches of the Gulf of Thailand and 92% and 64% of those from the Andaman Sea coast. Further, pair trawlers in the Gulf of Thailand took around 10% of both fish catches (Table 2.8).

Table 2.8   Percentage catches of sharks and rays according to fishing gear and zones in Philippines and Thailand (data from SEAFDEC 1988).

TYPE OF FISHERY AND GEAR	PHILIPPINES						THAILAND			
	LUZON		VISAYAS		MINDANAO		GULF		INDIAN OCEAN	
	SHARKS	RAYS	SHARKS	RAYS	SHARKS	RAYS	SHARKS	RAYS	SHARKS	RAYS
LARGE SCALE										
Purse seine	3	2	11	8	-	-	1	0	-	-
Trawl	30	6	1	23	-	-	12	10	-	0
Otter trawl	-	-	-	-	-	-	63	82	92	64
Gill net	-	-	-	-	-	-	22	1	4	-
Hook & line	2	-	-	-	-	-	-	-	-	-
Others	-	0	0	-	-	-	-	-	-	-
SMALL SCALE										
otter trawl	-	-	-	1	-	0	-	-	-	-
Gill/drift net	21	30	8	42	15	81	1	3	-	29
Hook/long line	38	42	76	22	57	7	0	4	4	7
Trap	-	7	-	3	0	1	-	-	-	-
others	6	12	3	4	28	10	-	-	-	-
TOTAL CATCH (mt)	1513	3132	1742	1924	3879	5689	3436	5963	408	1631

In the Gulf of Thailand, large-scale gillnets accounted for 22% of shark catches but only 1% of the rays. Purse seiners caught small catches of both fish. In the Andaman Sea small shark catches were taken by large-scale gill nets. Small-scale elasmobranch fisheries in Thai waters are relatively important for their catches of rays by gill nets in the Andaman Sea coast where they caught almost 30% of the local ray catches. Small catches (less than 1 to 7% of local catches) of both fishes are also taken in small-scale hook and line and longline fisheries in both coasts. In the Gulf of Thailand, small-scale gillnets take only small catches of sharks and rays.

The main fishing grounds for sharks and rays is the Gulf of Thailand. During 1976-1989 catches from the Gulf averaged 2955t/yr of sharks (28% of all elasmobranchs caught) and

4885t/yr of rays (46%) while the Andaman Sea produced only 1042t/yr of sharks (10%) and 1709t/yr of rays (16%). The trend of shark catches during this period showed a slight increase in the Gulf of Thailand and a decrease in the Andaman Sea. Ray catches from the Gulf of Thailand increased considerably and diminished in the Andaman Sea (Figure 2.26).

No recent stock assessments for the area are known. Studies from the early 1970's based on swept area estimates of the 1963 and 1966-1972 research cruises (Menasveta et al. 1973) indicated stock biomasses of 2880t for sharks, 4404t for rays and 1988t for rhinobatids in the whole Gulf of Thailand and an estimated 5000t potential yield for all elasmobranchs. The study identified large reductions in biomasses of rays over that period and concluded that elasmobranch stocks were "heavily exploited", if not overexploited. However, these estimates might have been too low as total Gulf catches of elasmobranchs from Thailand and Malaysia were 10 439t in 1977, 10 959t in 1978 and 7621 in 1979. They maintained a level of about 8000t/yr for another 6 years rising above 10 000t/yr in the late 1980's. Nevertheless, the reductions in catch rates (Pope, 1979) show that the stocks of both sharks and rays have declined dramatically in the area.

2.2.4.7 Malaysia

The elasmobranch fisheries of Malaysian and those of Philippines and Thailand are among the smallest in Asia. Catches of sharks and rays comprise only 2.46% of the world catch of this group. The development of the fishery in Malaysia showed a slow growth from 1961 to the current level of 15 000t/yr (Figure 2.2). Elasmobranchs currently represent 2.2% of the total catch of Malaysia. Rays are more important than sharks in the catches. SEAFDEC data indicate that from 1976-1991 rays represented, on average, 60% of the elasmobranch catch and sharks the remaining 40%. Catches of sharks showed overall a slight declining trend while ray catches increased, mainly from 1986-1991 (Figure 2.27). The main species in the ray catches are *Rhyncobatis djiddensis* (which together with other ray species is processed as "shark fin"), *Gymnura* spp. and *Dasyatis* spp. *Scoliodon sorrakowa*, *Chiloscyllium indicum* and *Sphyrna* spp. are the most common shark species caught (C. Phaik, pers. comm. 1992).

Elasmobranch caught in Malaysia are predominantly bycatch of trawl fisheries; only a small amount taken in directed fisheries. Almost 95% of the catches come from trawl fisheries while small-scale directed fisheries take the remaining 5%. Of the 16 822t of elasmobranchs caught by Malay fisheries in 1988, the great majority were taken by large-scale fisheries, of which trawl fisheries were the most important. In both coasts of Peninsular Malaysia and the Sabah coast, between 60% and 70% of the local shark catches were taken with trawls, while those of rays were in the order of 72-93%. Purse seines caught less than 1% of sharks in Peninsular Malaysia. In the waters of Sarawak, 70% of local ray catches came from large scale otter trawls, but this gear only contributed 30% of the shark catches. In this area, other large-scale gears accounted for less than 1% of catches of sharks and rays.

Malaysian small-scale fisheries for elasmobranchs are not as important as large-scale fisheries for their contribution to total elasmobranch catches. In Sarawak, during 1988, this sector took 70% of the local shark catches using mainly gill nets (54%), longlines and hook and line (15%) with traps making a very small contribution (Table 2.7). Rays taken by small-scale fisheries were caught by hook and lines and longlines (17%) and gillnets (11%); small catches were also taken with traps. For both coasts of Peninsular Malaysia and Sabah, small scale gill nets fisheries took between 15% and 28% of the shark catch while hook and line and longlines accounted for about 9% of the catch in Peninsular Malaysia and 25% in Sabah. Catches of rays from small-scale fisheries in Sabah and off the west coast of Peninsular Malaysia were taken

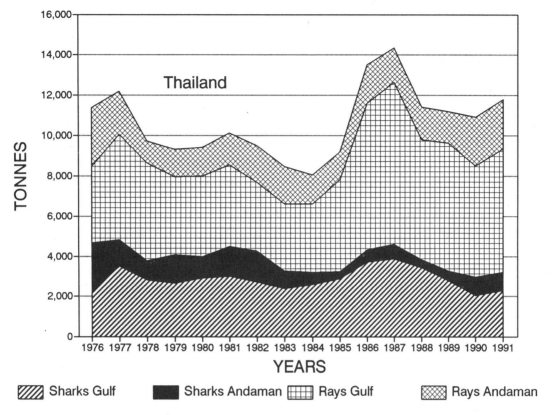

Figure 2.26. Elasmobranch catches of Thailand, by species groups and region, during 1976-1990 (Data from SEAFDEC).

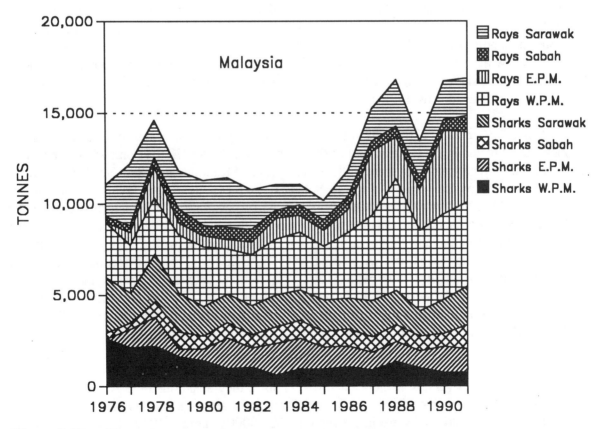

Figure 2.27. Elasmobranch catches of Malaysia, by species groups and region, during 1976-1990 (E.P.M.=eastern peninsular Malaysia, W.P.M.=western peninsular Malaysia) (Data from SEAFDEC).

mainly by hook and line and longlines and to a lesser extent by gillnets traps and other gear. The opposite occurred on the east coast of Peninsular Malaysia where most of the small contribution (5%) of small scale fisheries to the total rays catch came from gillnets.

As a consequence of the by catch of elasmobranchs, the most important fishing grounds are those of the trawl fishery - mainly peninsular Malaysia and Sarawak. During 1976-1989 sharks were taken mainly in Sarawak (1869t/yr or 15% of total elasmobranch catch), the west (1363t/yr or 11%) and east coasts of Peninsular Malaysia (1169t/yr or 9%) and in smaller quantities in Sabah (778t/yr, 6%). Sharks catches in these areas decreased in west Peninsular Malaysia but had relatively sustained yields in Sarawak and Sabah and were variable in east Peninsular Malaysia (Figure 2.27). For rays, the west coast of Peninsular Malaysia is the most important fishing area (3457t/yr, 28% of total elasmobranch catches) followed by Sarawak (2004t/yr, 16%) and the east coast of Peninsular Malaysia (1324t/yr, 11%), with Sabah contributing only 573t/yr (5%). The data show in increase in ray catches on both coasts of Peninsular Malaysia, relative stability in Sabah and strong variability in Sarawak. There are no existing management measures for elasmobranchs and the licence restrictions for trawlers only indirectly limit the catches, mainly those of rays.

2.2.4.8 Indonesia

Statistics for the elasmobranch fisheries of Indonesia were not recorded before 1971 but show a tremendous increase since the beginning of the FAO records. Indonesia holds the highest sustained rate of development for any elasmobranch fishery and currently has the largest fishery in the world. Indonesian catches amounted to almost 80 000t and there are no signs of levelling off (Figure 2.2). Indonesian fisheries represent 10.18% of the world's elasmobranch catch. Despite this, elasmobranchs are of only moderate importance in Indonesia, contributing 2.41% to Indonesian landings during 1987-1991. Contrary to most major elasmobranch fishing countries in the region, which harvest larger quantities of rays than of sharks or similar quantities of both, catches in Indonesia are dominated by sharks, which accounted for 66% of the average elasmobranch catches during 1976-1991.

SEAFDEC data (1976-1989) show that the most important areas for shark fishing in Indonesia are situated in the western part of the country, i.e. Java (9727t/yr on average and 21% of total elasmobranch yields), Sumatra (7837t/yr, 17%) and Kalimantan (5870t/yr, 12%) with the eastern provinces of Bali-Nusa Tengara, Sulawesi and Molluca-Irian Jaya, accounting for 1796t/yr (3.8%), 3157t/yr (7%) and 1983t/yr (4.2%) respectively. This pattern is similar for batoid catches except that Sumatra is the top producer with 6404t/yr (13% of total elasmobranch catches), followed by Java with 4670t/yr (11%) and Kalimantan with 2987t/yr (6%). In the eastern provinces Sulawesi is first with 1329t/yr (3%), Bali-Nusa Tengara second with 957t/yr (2%) and Molluca-Irian Jaya third with 518t/yr (1%). The catches of sharks and rays show increasing trends over the period in all provinces, except those of sharks in Molluca-Irian Jaya and both groups in Bali-Nusa Tengara which had rather poor development (Figure 2.28). These last two areas could be the most suitable for future increases in the fishery.

In addition to the Indonesian catches, large quantities of sharks have been harvested by the Taiwanese driftnet vessels in Indonesian waters since they abandoned the Australian EEZ in 1987. This fleet was capable of taking at least 7000t/yr of sharks and catches in the area between north Australia and Indonesia were in the region of 25 000t/yr before 1979 (Stevens, 1990). In the light of these combined catches, it is surprising that yields from Indonesia keep increasing annually. There are no apparent research or management programmes for elasmobranchs in Indonesia and the question of the potential of shark fisheries in the area becomes more intriguing as catches keep growing. Much attention should be paid to this fishery if catches are to be sustained.

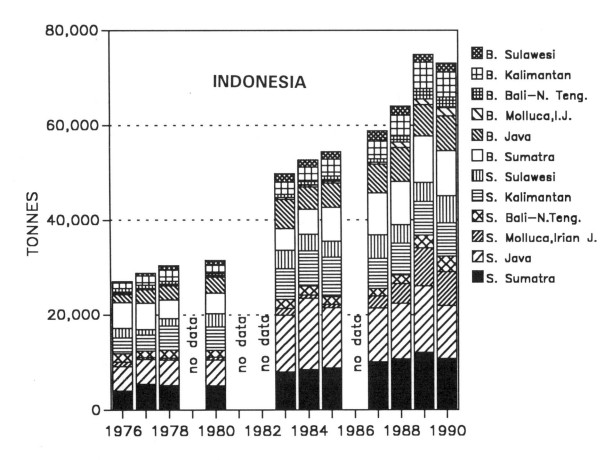

Figure 2.28. Elasmobranch catches of Indonesia, by species groups and region, during 1976-1990 (B=batoids, S=sharks) (Data from SEAFDEC).

2.2.5 Australian subcontinent

2.2.5.1 Australia

Elasmobranch fisheries in Australia are small and barely classifiable as "major fisheries" having only temporarily exceeded 10 000t/yr during the late 1980's (Figure 2.2). They only contribute 1.46% to the world elasmobranch catch (1987-1991). Nevertheless, Australian shark fisheries are among the most documented and managed elasmobranch fisheries in the world. This is probably directly related to the importance of elasmobranchs in the catches of Australian fisheries. FAO data for 1987-1991 show that elasmobranchs contribute 4.8% of the landings in Australia, the third highest percent importance in the world. Further, these are mature fisheries and form part of the fishing tradition of the country. Stevens (1990) reviews Australian shark fisheries and gives their history back to the end of the 19th century when fisheries for school sharks' liver oil and fins already existed in southeastern Australia.

FAO data are not presented by species or species groups and only the geographical distribution of the catches is discernible. The bulk of catches come from the Area 57 probably reflecting catches from the southern shark fishery for *Mustelus antarcticus* and *Galeorhinus galeus*. Small catches of elasmobranchs come from Area 81. Catches in Area 71 are negligible (Figure 2.29).

Historically, the most important elasmobranch fishery in Australia has been the southern shark fishery which provides the major part of the elasmobranch catches of the country. Walker *(1988), Anonymous (1989) and Stevens (1990) summarize the situation for this fishery.* School sharks, *(Galeorhinus galeus),* were the original targeted species at least since 1927, when records began to be taken regularly. Other important species in the fishery are the gummy shark (Mustelus antarcticus), the sawsharks (*Pristiophorus cirratus* and *P. nudipinnis)* and the elephant fish (*Callorhynchus millii).* Management of the fishery began in 1949 when a minimum size of 91cm TL was introduced for school sharks in Victoria. Protection of nursery areas in coastal lagoons followed later. The fishery expanded from coastal to offshore operations in the mid-1940's and catches gradually grew until 1969. The fishery suffered a temporary reduction in yields following the combined effect of the introduction of monofilament gillnets and a ban in Victoria of school sharks longer than 104cm TL due to impermissibly high mercury concentration in their flesh. The introduction of gillnets was intended to boost the decreasing catches of school sharks but this also brought about big bycatches of gummy sharks which had previously been regarded as undesirable species. Because of the size restrictions on school sharks the gummy sharks displaced school sharks as the main species in the catches. Soon, revised size limits allowed school sharks between 71-112cm TL to once more be taken in the Victorian fishery and total catches rose to a peak of 3754t (dressed weight) in 1986 with both species contributing approximately equally to the catch. Thereafter, catches slowly declined.

Most of the catch in the southern shark fishery is taken with monofilament gillnets and longlines but some catches is also taken by trawlers. Gillnets vary in size geographically and the mesh size ranges from 15cm (legal minimum) to 20.23cm with 17.78cm being most common. Gillnets used are typically 1.7m deep with a hanging coefficient of 0.6 (Kirkwood and Walker 1986). Gillnets are the main source of total shark catches (90% of the gummy shark and approximately 75% of the school shark catches). Longlines are typically 10km long and rigged with several hundreds of hooks. Although less important than gillnets their utilization has grown lately, especially in Tasmania. The most important fishing grounds for *Mustelus antarcticus* are primarily Bass Strait and secondarily in South Australia. The opposite is true for *Galeorhinus galeus* which, until recently, almost equalled those of the other areas in the Tasmanian catches. The contributions to the total shark catches of 1987 by gear and area are: Bass Strait, gillnets 47.3%, longlines 7.4%; South Australia, gillnets 27.3%, longlines 1.3%; Tasmania, gillnets 10.9%, longlines 10.4% (Anonymous 1989).

The fishery is a model for management of elasmobranch resources. Fishing effort has expanded in all areas and drops in gillnet CPUE (kg/km/hr) for both species have led scientists to suspect that both stocks are declining. As a result, a monitoring program and a special research group have been set up to study the fishery and several projects funded by the fishing industry and government agencies are being carried out. Their approach is comprehensive with research ranging from biological studies (Moulton et al. 1992) and the construction of databases and specific simulation models for the management of the fishery (Walker 1992, Sluckzanowski et al. 1993) to economic analyses (Campbell et al. 1991). The biology of the species is well known and suggests separate breeding populations for each species. However, concerns have been raised about the spatial structure and dynamics of the populations. Present investigations concentrate on the spatial dynamics of the stocks and the vulnerability of juvenile school sharks to commercial and sport fisheries in nursery areas of Tasmania. The recent concerns about overexploitation of the stocks has led to effort reductions by about a 50% through an elaborate licensing procedure. Longline effort was not considered in the scheme and this type of effort grew rapidly as a result of the restrictions imposed to gillnetters. It caused the overall effort reduction to fall short of that intended.

There is a smaller shark fishery operating in the lower western and south western coast of Western Australia. Catches are dominated by *Furgaleus macki* and *Mustelus antarcticus* but substantial catches of *Carcharhinus obscurus* are also taken (Lenanton et al 1990). Catches are about 1600t/yr and about 10% of the Australian catch of gummy shark comes from this fishery. Management measures include licence limitations, gear restrictions and a recent prohibition of shark fishing in waters from Shark Bay northward to North West Cape (Anonymous 1992).

The northern Australia shark fishery was started in 1974 by Taiwanese gillnetters exploiting sharks, tuna and mackerel in offshore areas of the Arafura sea. Taiwanese pair-trawlers fishing in the same areas also took sharks as by catch (see Section 2.2.4.4). Sharks comprise approximately 80% of the catch with 55% being *Carcharhinus tilstoni* and *C. sorrah*. At the beginning of the 1980's Australian fishermen became interested in these resources and small fisheries spread in inshore waters from the Northern Territory to the north of Western Australia and Queensland. Catch composition is similar to that of the offshore Taiwanese fishery and landings have fluctuated between 50 and 400t/yr (Stevens 1990). Although stocks declined due to overexploitation by the Taiwanese fleet, with the fleet move to Indonesia in 1987 the stocks are believed to be recovering. No management measures for the small domestic fishery are thought necessary at present. This fishery has been closely monitored and several research projects have been conducted by the Northern Territory Department of Primary Industry and Fisheries and the Commonwealth Scientific and Research Organisation (CSIRO).

The future development of a shark fisheries in North Australian waters is limited by high concentrations of mercury and selenium in most species of carcharhinids and sphyrnids. Lyle (1984) estimated that only 49% of the catch in weight could be retained if the maximum permitted level of mercury is 0.5 mg/kg. Further, market restrictions have prevented tropical catches from entering the main market for shark meat in Melbourne (Rohan 1981). Some recent arrangements have been made in the northern shark fishery to prevent overexploitation. Several restrictions have been introduced in different areas under Commonwealth jurisdiction since January 1992.

2.2.5.2 New Zealand

Elasmobranch fisheries in New Zealand remained under 10 000t/yr until recently. Although current catches are not much larger there has been an increasing trend since the late 1970's (Figure 2.2). Elasmobranch fisheries are moderately important for New Zealand with catches making 2.19% of the total fishery production during the last 5 years reported. New Zealand fisheries for sharks are another example of continuing research and management. On a global scale, these fisheries are small, contributing only 1.73% to world elasmobranch production (Table 2.2).

According to FAO data for 1977-1989 the yields of the different elasmobranch groups in New Zealand are quite variable. Dogfish (mostly *Squalus acanthias*) catches show a tremendous increase while catches of smoothhounds show a decline. Batoid and elephant fish catches grew moderately and the catch of grey sharks (mostly tope) greatly expanded and contracted during this period (Figure 2.30).

Recent information of the N.Z. Ministry of Agriculture and Fisheries shows that during 1989-1992, approximately 15% of the catch consisted of elephant fishes (*Callorhinchus milli*) and chimaeras (*Hidrolagus* spp.), 18% of tope shark (*Galeorhinus galeus*), 12.5% of rig (*Mustelus lenticulatus*), 33% of piked dogfish (*Squalus acanthias*), 17.5% of the skates *Raja nasuta* and *R. innominata*. The remaining 4% consisted of 13 species of large and deepwater sharks and at

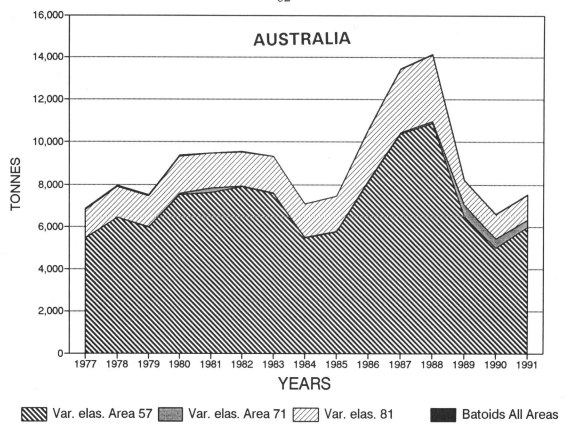

Figure 2.29. Elasmobranch catches of Australia, by FAO statistical areas, during 1977-1991 (Data from FAO).

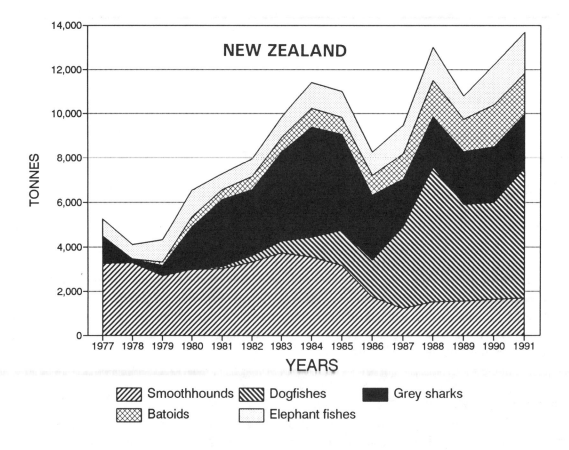

Figure 2.30. Elasmobranch catches of New Zealand, by species groups, during 1977-1991 (Data from FAO).

least 3 species of batoids. About 40% of the total is by catch of trawl fisheries while the other 60% is mainly taken directly with longlines and setnets. Elephant fishes are caught mainly off Canterbury; tope sharks and rigs are taken all around New Zealand.

Francis and Smith (1988) analyze the catches of rig around New Zealand and summarize some information about this fishery. The rig fishery is strongly seasonal concentrated during the austral spring and summer months. The catches are mostly exported to Australia. Almost 90% of the catches were taken as by catch of trawl fisheries during the mid 1960's, but the increase in demand and introduction of monofilament gillnets changed the pattern of exploitation and presently setnets account for 80% of the landings. Francis and Smith report that CPUE declined in three of the five zones analyzed during 1974-1985 and that in several areas stock sizes appear to be down to one third of their original sizes. Presumably, these are part of the reason for the imposition of management regulations in this fishery.

Management measures for the main elasmobranch species in New Zealand include TACs, a percentage of which go to ITQ holders. In 1992 the TACs were 636t for elephant fishes, 2070t for rig and 3087t for tope shark (Annala 1993). Catching basking sharks is prohibited and there are current proposals to include more elasmobranch species under the quota management system. Research in New Zealand has concentrated on rig and piked dogfish (Francis and Mace 1980, Hanchet 1988, Francis 1989, Massey and Francis 1989, Hanchet 1991, Francis and Francis 1993).

Some small quantities of livers from deep water squaloid sharks are currently utilized from the bycatches of the orange roughy (*Hoplostethus atlanticus*) deep trawl fisheries of New Zealand (King and Clark 1987), although large quantities of the sharks are also discarded at sea (see Section 2.3). Results from research cruises indicate that the stock of these deep sea sharks could sustain yields of no more than 2250t/y.

2.3 Bycatches and Discards of Elasmobranchs at Sea

Several large-scale fisheries operating in the high seas around the world are known to take a substantial bycatch of elasmobranchs, particularly sharks. Although sharks are retained and utilized in some of these fisheries, they usually are dumped, sometimes alive after their fins have been chopped off. The survival of released sharks varies depending on the type of gear used. Trawl and gill nets and perhaps purse seines, almost certainly cause 100% mortality. While longlines permit prolonged survival of sharks by allowing limited movement and thus some respiration, survival rates depend on the metabolism and endurance of individual species. Overall, it is believed that most of the bycatches of sharks in large-scale fisheries have high mortality. This might not be true for batoids which generally have different mobility requirements in order to respire. However, their catches are normally small in large-scale high seas fisheries due to their more demersal habits.

The amount of elasmobranchs killed in large-scale high seas fisheries is poorly known and has not been systematically assessed and an unknown part of the bycatch is discarded at sea. Reports on the sharks taken by the countries involved in these fisheries do not reflect the actual by catches but only the amounts retained. A purpose of this section is to present the available information on the most important large-scale fisheries of the world and evaluate the extent of their elasmobranch bycatches, the amounts taken and the total discards.

Until recently, there were two main large-scale fisheries catching and discarding significant numbers of elasmobranchs in their operations - driftnet and longline fisheries. Due to international pressures, and following **UN Resolution 44/225**, all large-scale driftnet fisheries were phased out of international waters at the end of 1992. They are discussed here because of the importance of their bycatches. In addition to longline and driftnet fisheries other large-scale fisheries with minor elasmobranch bycatches (tuna purse seine and pole and line fisheries) are briefly discussed. The deep trawl fisheries for orange roughy are also mentioned because of their potential impact on deep water shark populations. Attention is drawn to the assessments of the elasmobranchs caught and their catch rates. Incidental catches are estimated where no figures exist and are compared with reported landings for each fishery, or country, in order to assess the quantities of elasmobranchs wasted each year and which are not included in the official statistics of world fisheries.

2.3.1 Drift Gillnet Fisheries

For the last few decades, several countries, chiefly Japan, Korea and Taiwan (Prov. of China), prosecuted large-scale fisheries using drift gillnets on the high seas of many oceans. Typically, vessels deployed several kilometres of gillnet which efficiently trapped the relatively dispersed resources they aimed for. Unfortunately, they also captured many other non-target species, sometimes in vast quantities which commonly were not utilized. Concern over the impact of drift gillnets on the world's oceanic animals and ecosystems has been focused mainly on marine mammals, however, it is now known that sharks were among the most frequently caught non-target animals in some of these fisheries though little attention was paid to the effect of this gear on their populations. Although all large-scale driftnet fisheries on the high seas stopped as of December 1992, an assessment is attempted here of the effects on sharks and ray populations. Though most of this mortality has ceased, its effects may still effect subsequent generations of elasmobranchs.

The most important large-scale driftnet fisheries are examined to estimate the quantities and kinds of elasmobranchs that were caught in these global operations. The description of these fisheries is based on Northridge (1991) and bulletins of the International North Pacific Fisheries Commission (INPFC) (Myers et al. 1993, Ito et al. 1993) which give more detailed information.

2.3.1.1 North Pacific Ocean

Until recently, there were three main large-scale driftnet fisheries in the North Pacific, the salmon fishery, the flying squid fishery and the large-mesh fishery for tunas and billfishes. As a result of these fisheries the North Pacific was the most heavily exploited area in the world by driftnets. This was probably a consequence of the geographic location of the three large-scale countries involved in driftnetting.

Salmon fishery

The Japanese fleet was the largest in this fishery. Canadian and US fleet sizes are still considerable but they use small driftnets (< 500m per vessel) and fish exclusively in the coastal EEZ waters. There were two Japanese fisheries for salmon: (1) the mothership operation in international waters of the North Pacific south of the Aleutians and in the Bering Sea and, (2), the land based fishery in the high seas East of Japan (Figure 2.31). In general, during the past two decades the Japanese salmon fishery showed a consistent decline in effort that involved reductions in number of vessels, fishing area and fishing season.

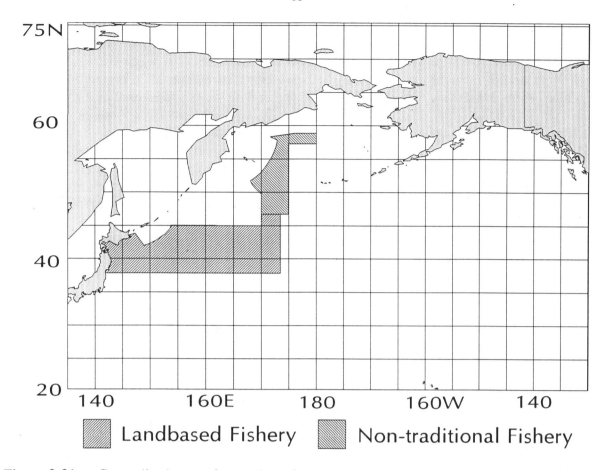

Figure 2.31. Generalized area of operation of the Japanese land-based and non-traditional (ex-mothership) fisheries in 1990. (Based on INPFC 1993).

Of these fisheries, the mothership processing ships supported some 40 smaller catcher vessels. The fishing grounds were divided in subareas with different opening and closing seasons, although the fishery season only ran from 31 May to 31 July. The fishery contracted its operations primarily due to pressure from the USA, Canada and the former USSR. During 1990 and 1991 operations were converted to a landbased fishery by eliminating the motherships. Catches peaked in 1956 when approximately 9 300 000 tans were set, only 238 700 tans were set in 1991, the last year of the fishery (FAJ, 1991). Tans are independent net panels which are the working unit of driftnets and are typically 45-50m long in the salmon fishery. Driftnets are 8-10m deep and are constructed of nylon monofilament with mesh sizes in the range of 121-130mm. Each vessel deploys a maximum of 15km of net in a dusk-to-dawn operation.

Two types of vessels were operated in the land based fishery: coastal boats of <30 GT and medium size vessels of 30-127 GT. Effort in this fishery also declined significantly latterly and the fishing area was reduced. There was a peak of 1400 coastal vessels in the mid-1970's but only 678 by 1978-1988 (Northridge 1991). There were 374 vessels over 30 GT in 374 boats in 1972; by 1991 there were only 83 (Myers et al. 1993). The number of sets peaked at approximately 19 700 in 1966 and declined to 4 100 (781 176 tans) in 1989 (FAJ, 1990) only 374 990 tans were set in 1991 (FAJ, 1991). During the last years of the fishery the season ran from late May to the end of June. Gillnets used by the landbased fishery were similar to those of the mothership fishery but with smaller mesh sizes (110-117mm). Coastal vessels of <10 GT set less than 10km of net per night while offshore vessels set up to 15 km.

Detailed reports on the by catches of non-target species in these fisheries (Northridge, 1991) are strongly biased towards studies dealing with marine mammals and birds; sharks are mentioned only as a side issue. However, the Fisheries Agency of Japan (FAJ, 1987, 1988, 1989) reported gillnet by catches of several non-target species in their research cruises for salmon. Table 2.9 show results for 1986-1988 together with the estimated total by catch of sharks taken in 1989 when 1 097 630 tans were set. Blue sharks are the most frequently reported shark species. The total by catch in the fishery for 1989 is estimated at 11 492 sharks consisting of 8 species, for approximately 108t. These estimates should be treated with caution. First, the areas and gear used in these research cruises appear to be different to those of the commercial fishery with at least some very small mesh sizes among the research driftnet effort. This will affect the catch rates of most species, both through changes in efficiency of the gear and availability of each species (e.g. blue sharks are not expected to be caught in the Bering Sea in high numbers due to their more temperate distribution). Direct extrapolation to the total fishery should be done carefully. Second, most of the catch rates of sharks reported in Table 2.9 seem too low compared with other studies.

Table 2.9. Estimation of shark bycatches in the Japanese salmon fisheries, based on information from research cuises.

Species	1986 (24,549 tans)	1987 (17,056 tans)	1988 (17,805 tans)	Catch Rate a) (sharks/1000km)	Estimated Numbers in Catch 1989 b)			Likely weight (kg) c)	
					Landbased	Mothership	Total	per shark	in the catch
Unid. Lamnidae	0	1	2	1.01	39	16	55	50	2,771
Lamna ditropis	25	26	23	24.91	973	394	1,367	50	68,359
Isurus oxyrinchus	13	1	2	5.39	210	85	296	50	14,780
Prionace glauca	142	188	79	137.69	5,378	2,179	7,556	2.42	18,287 d)
Squalus acanthias	73	33	8	38.38	1,499	607	2,106	2	4,212
Isistius brasiliensis	1	1	0	0.67	26	11	37	0.75	28
Mustelus manazo	1	0	2	1.01	39	16	55	2	111
Triakis scyllium	0	0	1	0.34	13	5	18	2	37
Totals	255	250	117	209.39	8,179	3,313	11,492	159.17	108,586

a) assuming 50m tans in research cruises
b) based in effort reported by FAJ (1990)
c) Considering sizes expected for 110-130 mm mesh
d) Calculated from LeBrasseur et al. (1987) length frequency data, Pratt (1979) TL-FL relationship, and Strasburg (1958) L-W relationship.

Although no other direct reports for the salmon fishery were found except those of the FAJ research cruises, results from Canadian research cruises (LeBrasseur et al. 1987) can be used to derive alternative catch rates for sharks. The results obtained for blue and salmon sharks are of an order of magnitude higher than those calculated from FAJ data. They give values of 5275 and 194 sharks/1000 km of net respectively (Table 2.10). These research cruises were designed to assess the salmon by catches of the squid fishery but employed nets nearly identical to those of the commercial salmon fishery. Thus, their results should more accurately reflect the catch rates of sharks in the commercial salmon fishery.

In general terms, the total catches of sharks in the Japanese salmon fisheries is believed to have been relatively small compared with other driftnet fisheries in the north Pacific. Even considering the alternative catch rates of 5502 sharks per 1000km of driftnet derived above, some

Table 2.10. Alternative estimates of shark bycatches in Japanese salmon fishery based on Canadian research cruise (LeBrasseur et al. 1987).

Species	Sharks caught (618 tans)	Catch rate per/1000km of net	Estimated numbers in 1989 catch	Likely weight (kg) per shark	Likely weight (kg) in 1989 fishery	
Prionace glauca	163	5,275	289,504	2.42	700,601	a)
Lamna ditropis	6	194	10,657	50	532,830	
Squalus acanthias	1	32	1,776	2	3,552	
Total	170	5,502	301,937	54.42	1,236,983	

a) Calculated from LeBrasseur et al. (1987) length frequency data, Pratt (1979) TL-FL relationship, and Strasburg (1958) L-W relationship.

300 000 individuals, or approximately 1237t, are estimated to have been caught during the 1989 season in this fishery. This relatively small catch is mainly a function of the size of the fishery, which has contracted year-by-year. As a reference point, according to Shimada and Nakano (1992), some 34 000 large and adult salmon sharks were landed from the salmon driftnet fishery in Japan in 1960. Further, reports for the early 1980's (Paust 1987) indicate that 25 000 salmon sharks (*Lamna ditropis*) were taken each year by the Japanese salmon fishermen in the central Aleutian region. Considering pertinent effort statistics and the catch rates obtained from LeBrasseur et al. (1987), a total of less than 1600 salmon sharks are thought to have been taken in the area south of the Aleutians in 1989, i.e., a reduction of about 95% in salmon shark mortality accompanied the decline of the fishery. Although there is not enough information to assess the level of catches and discards of sharks that took place in this fishery, it is possible that some of the salmon sharks would have been kept and utilized. This is suggested by reports of specific fisheries for this species taking place in NE Japanese waters off the Oyashio front (Paust 1987, Anon. 1988) which indicate that the salmon shark is sought by Japanese fishermen. However, the incentive to keep salmon sharks probably should be weighted against the availability of space and danger of spoilage of the valuable salmon catch. In July 1991, all Japanese salmon driftnet fisheries in the high seas ceased. Most of the fleet was disbanded although a minor part moved to Russian EEZ waters under a Russian joint-venture. There are no data available about this new salmon driftnet fishery but judging from the calculations made above its bycatches of elasmobranchs should be minor.

Flying squid fishery

In the late 1970's, a major driftnet fishery for flying squid (*Ommastrephes bartrami*) was started in the Central North Pacific by Japan (in 1978), then Korea and Taiwan (Prov. of China). In 1990 almost 740 vessels from these countries prosecuted this fishery. Yatsu et al. (1993) summarize most of the information available for Japan. Japan limited the number of vessels and the area open to this fishery (Figure 2.32) by a northern boundary which moved through the year to avoid taking salmon - which was prohibited by the flying squid fishery. Japanese vessels were classified in two categories: 60-100 GT and 100-500 GT. The fishing season for these vessels ran from June 1st to December 31st, although two types of licences, for 7 and 4 months, were issued within the season. Nylon monofilament (0.5mm) driftnets were used with a mesh sizes of 100-135mm; 115-120mm sizes were the most common. Tans were 9-10m deep and 33-42m long. Each vessel set between 15 and 50km of net although some reports indicate that the most common sets were close to 50km.

Korea joined the fishery in 1979 (see Gong et al. 1993 for a full account). Korean squid driftnet vessels were mostly about 350 GT, but some exceeded 400 GT. The Korean fleet fished from April to early August in an area partially overlapping the Japanese grounds and from early August to mid-December for smaller squid east of Japan (Figure 2.32). Their driftnets had 50m tans with mesh sizes of 76-155mm. In the first fishing area mesh sizes used were 105-115mm while those used in the grounds east of Japan were 86-96mm. According to Gong et al. (1993) Korean vessels deployed about 28km of driftnet in the early 1980's but increased to 45km in 1990.

Information on the Taiwanese squid fishery, which started in 1980, is scarce and most of the information here is based on the brief account of Yeh and Tung (1993). Vessel sizes ranged from 100-700 GRT but most were 200-300 GRT. Driftnetters larger than 400 GRT were introduced mainly in 1984 while those larger than 600 GRT entered during the 1986-1987 season. Taiwanese driftnets for squid are believed to have been constructed of monofilament nylon. Mesh sizes ranged from 76-120mm with each tan measuring between 15 and 40m in length. Typical total lengths of driftnet deployed per boat were 31-41km (Fitzgerald et al. 1993). Taiwanese vessels were allowed to fish year round (Pella et al. 1993) but the fishing season was apparently from June to November (Yeh and Tung 1993) in an area similar to that fished by Korea but west of the Japanese EEZ (Figure 2.32).

Effort statistics for these fisheries have only recently been available. According to data provided by Yatsu et al. (1993), Gong et al. (1993) and Yeh and Tung (1993) the total number of vessels from the three countries prosecuting the squid driftnet fishery during 1988-1990 was 792, 784 and 737 respectively. Statistics on the total number of tans deployed by Japan and Korea are also available. Unfortunately, Taiwanese statistics do not separate effort between the squid fishery and the large-mesh driftnet fishery as their boats carried both gears and deployed either depending on the expected catch. Further, Taiwanese effort statistics are given only in total "vessel-days" fished (Table 2.11).

The total number of standardized tans set by the Taiwanese fleet in the squid fishery can be estimated using comparative data on total length of sets for vessels from each county. Fitzgerald et al. (1993) estimate a total of 51-61 km of driftnet per Japanese vessel and 31-41km per Taiwanese vessel. Data from Yatsu et al. (1993) indicate that Japanese vessels deployed an average of 997.43 tans (50m each) per fishing day during 1989 and 1990. The effort of Taiwanese vessels is assumed to be allocated equally to the flying neon squid and the large-mesh fisheries. Assuming the number of tans per vessel is equal in the Japanese and Taiwanese fleets, total estimated effort was 4 471 678, 5 616 888 and 3 595 855 standardized (50 m) tans for the Taiwanese fleet in the squid fishery for the years 1988-1990 respectively. Total effort for the three countries in this fishery is estimated at 64 782 236 tans (3 239 112 km) for 1989 and 50 922 388 tans (2 546 119 km) for 1990.

There are several sources of information on catches of non-target species, chiefly in the form of research cruises and more recently from observer programmes. Results from some research surveys enable an assessment of catch rates in numbers for blue, salmon and four other species of sharks, size structure and catch rate in kg/m for blue sharks, percentage distribution by mesh size for blue and unspecified shark species and differences in blue shark catches between surface and subsurface squid driftnets (FAJ 1983, Murata and Shingu 1985, Murata 1986, 1987, Rowlett 1988, Murata et al. 1989, Yatsu 1989, Ito et al. 1990). However, results from these surveys suffer the same problems as for the salmon fishery research surveys. Japanese and Korean research cruises use a variety of mesh sizes which extended above and below the size of those used by the commercial fishery. The results therefore are limited to assessing total catches of non-target species. More useful information comes from the observer programmes on commercial vessels. Data from Japanese observers for 1988 (FAJ 1989) give catch rates of 536 blue sharks per 1000km of net. However, collective data from Japanese, Canadian and U.S. observers for 1989 (Gjernes et al. 1990) report 814 blue sharks per 1000 km of net.

Table 2.11. Effort statistics for the flying squid driftnet fishery in the North Pacific for the period 1988-1990 (from Yatsu et al. 1993, Gong et al. 1993 and Yeh & Tung 1993).

Year		Japan	Korea	Taiwan	Total
1988	# boats	463	150	179	792
	days fished	-	-	14,010	
	total tans	36,055,567	24,594,370	-	
1989	# boats	460	157	167	784
	days fished	33,646	-	17,598	
	total tans	34,385,032	24,780,316	-	
1990	# boats	457	142	138	737
	days fished	23,656	-	11,266	
	total tans	22,769,857	24,556,676	-	

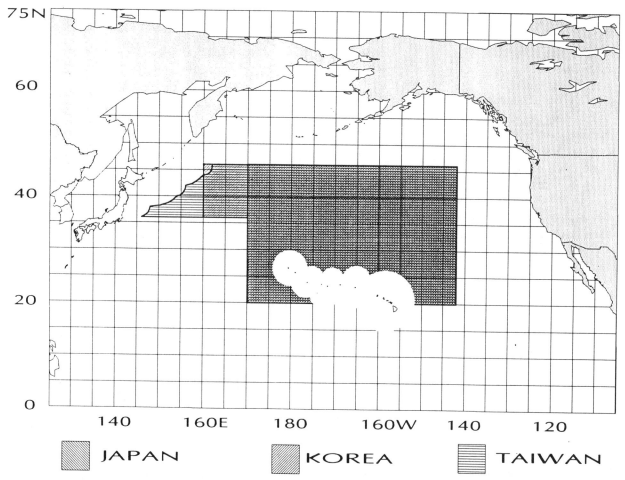

Figure 2.32. Legal boundaries of the Japanese, Korean and Taiwanese flying squid driftnet fisheries. (Redrawn from Pella et al. 1993).

Data for the 1990 observer programme (INPFC 1991) are more detailed and show that 12 elasmobranch species were taken as bycatch in the fishery. The catch rates for blue sharks was 718/1000 km of driftnet, followed by salmon sharks, 55/1000km of driftnet. Other large shark species caught, perhaps by entangling, were the thresher shark (*Alopias vulpinus*), shortfin mako (*Isurus oxyrinchus*), great white (*Carcharodon carcharias*) and basking shark (*Cetorhinus maximus*) (Table 2.12). Observer data from the Korean fleet for 1990 give an estimated catch rate of 32.08 sharks and rays per 1000 poks (Korean tans), equivalent to 641.6/1000km of net. This is slightly low, compared to the 785/1000km of net estimated for the Japanese fishery. Data on fishes in most of the observer programmes for the North Pacific driftnet fisheries are likely to be slightly underestimated. Only decked animals are counted thus unknown numbers of "dropoff" fishes are not included in the records. Despite this, observer programmes provide the best available information.

Table 2.12. Estimation of bycatches of elasmobranches in 1990 Squid driftnet fishery based on reports of observer programme on board commercial vessels (NPFC 1991).

Species	Numbers observed (2,281,896 tans)	Catch rate per/1000km of net	Numbers in total catch	Likely mean weight (kg)*	Weight in total catch (kg)
Unidentified shark	1,191	10	26,578	15?	398,672
Prionace glauca	81,956	718	1,828,915	7 (1)	12,802,407
Lamna ditropis	6,263	55	139,764	38.7 (1)	5,408,866
Isurus oxyrinchus	71	0.622	1,584	40	63,377
Alopias vulpinus	48	0.421	1,071	40	42,846
Squalus acanthias	8	0.070	179	2	357
Carcharodon carcharias	7	0.061	156	50	7,811
Isistius brasiliensis	5	0.044	112	0.75	84
Euprotomicrus bispinatus	1	0.009	22	0.20	4
Cetorhinus maximus	1	0.009	22	500	11,158
Dasyatis violacea	8	0.070	179	10?	1,785
Dasyatis brevis	1	0.009	22	10?	223
Unidentified ray	8	0.070	179	10?	1,785
Totals	89,568	785	1,998,783	-	18,739,376

* considering sizes expected for 100-135 mm. mesh
(1) from Yatsu et al. 1993

There are some estimates for elasmobranch by catches in the squid driftnets. Yatsu et al. (1993) estimate a total incidental catch of 723 933 blue sharks, 56 029 salmon sharks and 11 322 various sharks and rays for the Japanese fleet during 1990, making an estimated 7415t. Yatsu and Hayase's estimate considers sources of variability for cruises and sets sampled. However, their estimates of blue shark by catch for 1989 are almost double those for 1990 highlighting the variability in estimates and the changes in fishing effort during the period. Wetherall and Seki (1992) used a stratified estimate to obtain a total of 1.2-1.4 million blue sharks for the Japanese fishery during 1989 while Northridge (1991) estimated the total catch of blue sharks for the entire flying squid fishery during 1989 at 2.44 million individuals assuming the same effort level of 1988 and catch rates derived from Gjernes et al. (1990).

Using the effort statistics for 1990 and the results from the Japan-USA observer programme (INPFC 1991) the number of elasmobranchs caught by species and the estimated weight of their catches in the 1990 are summarized in Table 2.12. About 2 million sharks, an estimated 18 739t, were taken in the fishery. Of these, about 12 802t, or 1.8 million individuals, would be young blue sharks, which according to Nakano and Watanabe (1992), correspond to sharks 1-2 years old. Unless otherwise stated, the estimates of weight for each species are based on approximations made by the author that consider the relatively small mesh size of the nets and might be biased towards small sizes. These results are a minimum estimate of the catch of elasmobranchs by the fishery. Of the 18 700t of sharks caught, 8 400t would have been taken by Japan; Korea and Taiwan (Prov. of China) would have caught 9000t and 1300t respectively.

A great proportion of the elasmobranch bycatches are apparently dumped to the sea. Assessment of shark catches for the Japanese fleet in 1989 using the same procedure gave estimates of 1 800 000 individuals with a total weight of 12 654t. The reported catch of sharks by the squid fleet of Japan during 1989 is 237 734 individuals (FAJ 1990). If this figure is equal to the landed catch of sharks, about 1 560 000 sharks weighing some 10 900t were wasted in the operation. Some of the almost 95 000 salmon sharks estimated to be caught in the fishery might have been used as this species is more valued in Japan. An appraisal of the amount of elasmobranchs actually discarded by the fleets of Taiwan (Prov. of China) and Korea is not possible due to the lack of information on their landings from the squid fishery. The total catch of elasmobranch for the Korean and Taiwanese fleets in 1989, estimated in the same manner, is 9120t and 2067t respectively.

These estimates of elasmobranch by catches are approximate due to limitations of the available data. But they do highlight the problems in determining the size of the elasmobranch by catch and the proportions dumped at sea, e.g., estimates of total weight of the bycatch are sensitive to the average weights for each species used in the calculations. This is particularly true for blue shark which accounts for most of the by catches by number. Yatsu et al. (1993) use an average weight of 7kg for blue sharks but alternative calculations give an average of 2.4 kg/shark. This was estimated from the length frequency reports for blue sharks of LeBrasseur et al. (1987) and morphometric equations for the species provided by Strasburg (1958) and Pratt (1979). The estimate of 2.4 kg/shark is consistent with the results of Bernard (1986), Mckinnell et al. (1989) and Murata et al. (1989) for nets with the same characteristics as those of the commercial squid fishery.

The results derived here appear to be slight overestimates compared with other results. However, considering that observer programmes do not consider "dropoffs" from the nets, the present estimates may be closer to the real mortality caused by the driftnets and serve as an indication of the order of magnitude of the problem. Following this reasoning, previous appraisals of blue shark catches in the whole fishery (Anon. 1988) seem to be highly overestimated.

Efforts to minimize the take of non-target species in the squid driftnet fishery were unsuccessful. Data summarized by Gong et al. (1993) for Korean research experiments shows that shark bycatches can drop by up to 41% when subsurface driftnets are used instead of normal (surface) driftnets. Unfortunately, catches of the target species (neon flying squid) dropped by 73%, probably making operations with the subsurface driftnets unprofitable. As a result of international agreements the squid driftnet fishery of the North Pacific stopped at the end of 1992.

Large-mesh Driftnet Fishery

A large-mesh driftnet fishery for skipjack, marlin, albacore and other tunas on the high seas of the North Pacific was started in the early 1970's by Japan. However, this fishery ended on 31 December 1992 together with all other high seas driftnet fisheries in the area. It started in the coastal Japanese bluefin tuna fishery of the 1840's but by the late 1980's it covered an area extending from 140°E to 145°W (Figure 2.33). The fishing grounds were divided into two areas: a southern area open to fishing year round and a northern area with portions closed to fishing during some months to avoid catching salmonids. Recent reports indicate this fishery operated with vessels in the 100-500 GT range. Nets were made of nylon monofilament twines of 1.2mm diameter for smaller meshed nets and multifilament and multistrand for larger meshed ones. Mesh size was greater than 150mm. Meshes as small as 113mm have been recorded though most driftnets used 180mm (INPFC 1992). Tans were commonly 33-36 m long. Japanese boats were restricted to a maximum of 12km of net at a time. Recent figures show that 459 vessels from Japan participated in the large-mesh driftnet fishery in 1988 catching approximately 40 000t. Taiwanese vessels also participated in this fishery, but information is scarce. Apparently, up to 123 vessels from Taiwan (Prov. of China) took part in this fishery during 1989. The Taiwanese fish chiefly from June to December.

According to the most recent data (Fitzgerald et al. 1993), Japanese vessels deployed a total of 4 682 630 standard (50m) tans in this fishery during 1990. Taiwanese effort is assumed to be the same for that estimated for the squid driftnet fishery (see above) due to the combined nature of these fisheries (Yeh and Tung 1993). The total effort of both countries during 1990 was equal to a total of 413 924 km of large mesh driftnet.

Information on the kinds and numbers of elasmobranchs caught in this fishery has become available through the reports of the international observers programme (INPFC 1992). Catch rates and estimates of the total catches of sharks and batoids based on effort levels reported for 1990 indicate that about 150 000 sharks, equivalent to 1722t, were taken as by catch (Table 2.13). The average weights of some species were obtained from research cruises that used driftnets with mesh sizes 150-180mm (FAJ 1983) while others are estimated from the mesh sizes used.

The estimated elasmobranch by catch of 366 fish/1000km for the large-mesh driftnets is about half that of the squid fishery. This difference is related to the different selectivity of the different nets. Larger meshes allow a greater escapement of small non-target species. Blue shark catch rates are less than half of those observed in squid driftnets and catch rates for salmon sharks are even lower, though the average size of each species tends to be larger in the large-mesh fishery. Of the total estimated catch of elasmobranchs in this fishery in 1990, approximately 974t would have been taken by Japan and 748t by Taiwan (Prov. of China).

No estimates of past elasmobranch by catch could be found for this fishery to compare with the present values. Further, there are no data on the amounts of elasmobranchs landed from the large-mesh fishery in Japan or Taiwan (Prov. of China). Judging from the trends in other high seas fisheries, it is likely that most bycatches of sharks were discarded at sea.

2.3.1.2 South Pacific Ocean

Large-scale driftnet fishing stopped in 1991 in the South Pacific. Previously Japan and Taiwan (Prov. of China) fished chiefly for albacore with large-mesh driftnets (Northridge 1991). Due to pressure from coastal states in the area it was agreed to terminate these high seas South

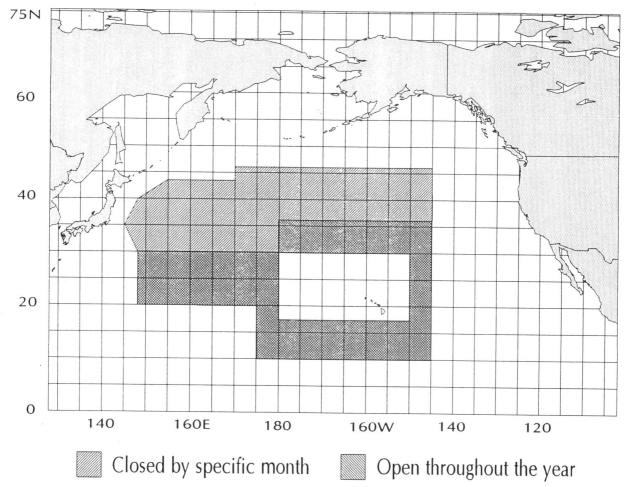

Figure 2.33. Area of operation of Japanese large-mesh driftnet fishery (Redrawn from Nakano et al. 1993).

Table 2.13. Estimated bycatches of elasmobranchs in the 1990 North Pacific large-mesh driftnet based on repors of the observer programme 1990 (I.N.P.F.C. 1992).

Species	Numbers observed (513,367 Tans)	Catch rate per 1000km of net	Numbers in Total catch	Likely mean Weight (kg)	Weight in Total catch (kg)
Unidentified shark	57	12.00	4,967	25 ?	124,177
Prionace glauca	7,692	300	124,040	9.2 (1)	1,141,168
Lamna ditropis	136	5.30	2,193	32.5 (1)	71,276
Isurus oxyrinchus	592	23	9,547	30 ?	286,395
Alopias vulpinus	6	0.23	97	167 (1)	16,158
Squalus acanthias	1	0.04	16	2.5	40
Carcharodon carcharias	35	1.36	564	47.7 (1)	26,922
Isistius brasiliensis	305	12	4,918	0.85	4,181
Euprotomicrus bispinatus	156	6.08	2,516	0.25	629
Cetorhinus maximus	2	0.08	32	550	17,738
Triakidae	3	0.12	48	3	145
Sphyrnidae	2	0.08	32	127 (1)	4,096
Dasyatis violacea	73	2.84	1,177	12 ?	14,126
Dasyatis brevis	8	0.31	129	12 ?	1,548
Unidentified ray	69	2.69	1,113	12 ?	13,352
Totals	9,137	366	151,390	-	1,721,953

(1) Derived from F.A.J. (1983).

Pacific fisheries by 1991. It is unclear if the agreement relates only to the waters of the South Pacific Commission (Figure 2.34) or if it also includes the eastern waters of the South Pacific. Japan stopped all large-scale driftnet fishing in the area in 1990 (Nagao et al. 1993). No information on Taiwanese vessels activities is available however, it appears that elasmobranch bycatch in large-scale driftnets in the South Pacific should at present be little or nothing, even if vessels from Taiwan (Prov. of China) continue to fish there.

Some reports of elasmobranch catch rates in the South Pacific are given in Table 2.14 based on data from Sharples et al. (1990) and Watanabe (1990). Their sources of information are two research cruises conducted in the Tasman Sea and the Sub-Tropical Convergence Zone (STCZ) to the east of New Zealand between 30° and 45°S. Catch rates estimated from these data are 181 and 158 sharks/1000km of net for the STCZ and the Tasman Sea respectively, or 5035 kg/1000km of net for the Tasman Sea. While total elasmobranch catch rates seem relatively similar among both areas, strong differences in catch rates for individual species are evident from the detailed information (e.g. blue sharks are more frequently caught in the STCZ than in the Tasman Sea while the opposite is true for mako sharks). Further, the catch rate for the Tasman Sea is high compared to data given by Coffey and Grace (1990). These differences illustrate the difficulties faced in extrapolating from catch rates to total bycatches when the data are based on information limited to a particular area, fishery or season.

Table 2.14. Reported bycatches of elasmobranchs in South Pacific driftnet fisheries.

	STCZ (464 km of net)*		TASMAN SEA (766 km of net)**			
Species	Numbers Caught	Catch rate (#/1000 km)	Numbers Caught	Catch rate (#/1000 km)	Mean Weight	Catch rate (kg/1000 km)
Cetorhinus maximus	-	-	1	1.31	-	-
Prionace glauca	70	150.86	22	28.72	70	2,001
Lamna nasus	-	-	3	3.92	-	-
Isurus oxyrinchus	10	21.55	66	86.16	31	2,663
Isistius brasiliensis	-	-	10	13.05	-	-
Sphyrna zygaena	-	-	3	3.92	95	371
Dasyatis violacea	4	8.62	16	20.89	-	-
Total	84	181.03	121	157.96	195	5,035

* Data from Watanabe (1990)

** Data from Sharples et al. (1990)

Based on commercial vessel activities, Coffey and Grace (1990) estimated catch rates of 48 sharks/1000 km of net and a total bycatch of 3500 sharks from the Tasman Sea area for the 1990 season. Murray (1990) compiled data from several sources and provides information on percentage by weight of sharks in total catches of Japanese research campaigns using 3 types of driftnets along with total effort for each type of net. Shark catch rates calculated using this information, and assuming 50m tans are 16 362 kg/1000 km, for albacore nets; 14 618 kg/1000km, for slender tuna nets; and 21 781 kg/1000 km, for pomfret nets. Given the lack of estimates of the total amount of nets deployed in these fisheries it is necessary to use the above percentages of sharks as a bycatch for the albacore nets and the reported albacore catches for driftnet fleets in 1989 provided by Lawson (1991). Thus, estimates of total shark bycatches are: Japan, 3462t, Korea, 48t and Taiwan (Prov. of China), 2871t for a total of 6381t. This corresponds to the reported peak in albacore driftnet catches. Therefore, total by catch levels should have been smaller in the earlier and later years of this fishery. These estimated catches are for the waters of the South Pacific Commission (Figure 2.34) only and are crude estimates limited by the available information. Further, it is unknown if the data cited by Murray (1990) on which the by catch percentages are based, contain information from the whole South Pacific region or only part. Geographical variations in abundance are likely to considerably affect the by catch estimates. Without information about driftnetting activities in the rest of the South Pacific Ocean it is possible that, given the proportion of the South Pacific covered by the SPC (about ⅔), the by-catch of elasmobranchs in the whole Southern Pacific could have been 50% more than that calculated for the SPC zone, or a total of 9572t. There is uncertainty about this estimate, which is about half the driftnet catch of elasmobranchs in the North Pacific Ocean.

2.3.1.3 Indian Ocean

Several countries have extensive driftnet fisheries in the Indian Ocean but most coastal states, e.g., India, Pakistan and Sri Lanka, only fish inshore waters in small to medium-scale fisheries (Section 2.2.). The elasmobranch catches of these coastal states are assumed to be landed and reported in FAO statistics. Taiwan (Prov. of China) is the only country known to have large-scale driftnet fisheries in the international waters of the Indian Ocean but only limited information is available on their activities. Their tuna fishery started with one boat in 1983 and increased to 139 vessels by 1988. Fishing apparently occurs from November to March with driftnets of 200-220mm mesh size, 20-24m depth with 20-25 or 37-47km of net deployed per vessel. Fishing occurs in the North West and South Central Indian Ocean. Hsu and Liu (1991) report sharks to be 23 by number and 29% by weight of the total catches for the 1986-1987 season while for 1987-1988 season their contribution decreased to 0.52% and 2.07%. As no significant changes in the fishing area were observed between both fishing seasons, this reduction in shark bycatches is most likely caused by changes in discard practices. By multiplying the percentage composition of sharks by the reported total landings of 18 281t in the 1986-1987 season (IPTP 1990), 5405t of sharks are estimated to have been caught by the fishery. A total of 6108t of shark is estimated to have been caught during the 1988-1989 season assuming that the number of vessels increased by 13% over the 1986-1987 level.

2.3.1.4 Atlantic Ocean

Until recently, the only known large-scale driftnet fisheries in the Atlantic were a French albacore fishery and an Italian swordfish fishery. However, Taiwanese driftnet vessels were also believed to operate in this ocean during the early 1990's. Many other gillnet fisheries exist in the Atlantic and Mediterranean and in many cases large quantities of nets are deployed nightly. However, most of these fisheries are limited to coastal waters and are not within the scope of this section. A summary of these smaller fisheries is given by Northridge (1991).

The French albacore fishery began in the Bay of Biscay in 1986; 37 vessels operated in 1989. These boats troll during the day and use gillnets at night. Fishing occurs from June to September and extends from the Azores north and eastward, following the albacore. Nets are 20-36m deep with 80-120mm mesh size; a mesh size of 90mm is the most successful. While French reports indicate driftnet lengths of 2-6km per vessel, Greenpeace claims they are up to 20km long. The only available information on shark by catches indicates they were of the order of 6-10%. Woodley and Earle (1991) observed several French boats and report sharks (mostly *Prionace glauca)* as the most common by catch, amounting to 6.2% of the albacore catch. The

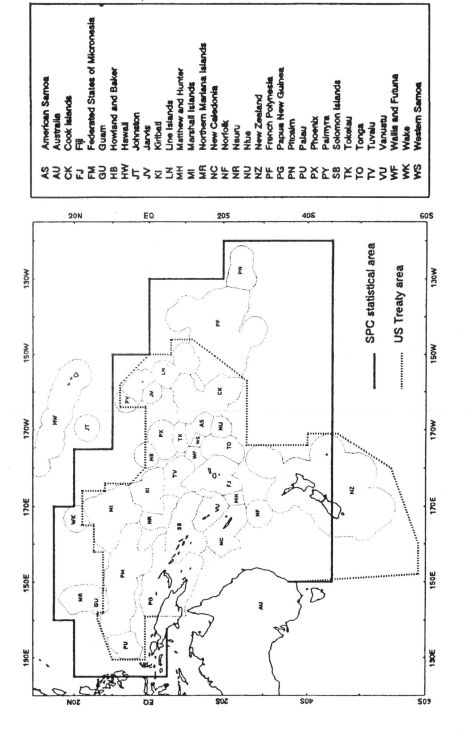

Figure 2.34. South Pacific Commission statistical area (Taken from Lawson 1991).

sharks caught ranged between 40-250 cm but were most common between 125-200cm. Woodley and Earle estimate catch rates of 1750 to 3520 sharks/1000km of net (including dropouts) for a total catch of 22 015 to 44 282 sharks during the 1991 French albacore fishery. This is equivalent to 430-865t of sharks assuming a mean total length of 175 cm for blue sharks. They reported a discard of 2 sharks at sea but no further information is available on the disposition of the shark by catches in this fishery. These shark catches could be included in the reported "various elasmobranchs" of France which amount to almost 10 000t/yr.

The use of driftnets in Italian fisheries for tuna and swordfish has a long history, but the fishery expanded only from the 1980's as a consequence of government support. According to Northridge (1991), this was one of the largest driftnet fisheries in the world before it was banned. By 1989, 700 vessels participated, 90% of them used nets of 12-13 km in length with depths of 28-32 m with mesh sizes of 180-400mm. A few vessels used less than 6km of net and a few others, more than 20km. The fishery pursued albacore and swordfish from Sicily and Calabria to the Ligurian Sea. While no information on catch rates of non-target species exists, several elasmobranchs have been reported caught by this fishery. Species commonly caught include thresher, blue and porbeagle sharks as well as manta and common eagle rays. Another three sharks are reported to be infrequently taken and 10 more as occasional taken species (Table 2.15). It is unknown if most of the catches were kept or discarded. It is impossible to estimate the amount of the total catch from available information, however, a large increase in landings of smoothhounds took place concurrently with the expansion of the driftnet fishery and it is known that other sharks are commonly merchandised locally as smoothhounds (De Metrio et al. 1984). It is possible that a considerable part of the shark by catch was landed by this fishery. Recent reports suggest that there are still some driftnetters in the Ligurian Sea using gear lengths above the permitted 2.5km per vessel for this area (ICCAT 1993a).

Table 2.15. Elasmobranchs caught in Mediterranean driftnets (adapted from Northridge 1991).

Common Name	Scientific Name
Species commonly caught	
Thresher shark	*Alopias vulpinus*
Blue shark	*Prionace glauca*
Porbeagle	*Lamna nasus*
Manta ray	*Mobula mobular*
Common eagle ray	*Mylobatis aquila*
Infrequent species	
Basking shark	*Cetorhinus maximus*
Shortfin mako	*Isurus oxyrinchus*
Smooth hammerhead	*Sphyrna zygaena*
Occasional species	
Bigeye thresher	*Alopias superciliosus*
Spinner shark	*Carcharhinus brevipinna*
Blacktip shark	*C. limbatus*
Dusky shark	*C. obscurus*
Sandbar shark	*C. plumbeus*
Great white shark	*Carcharodon carcharias*
Sharpnose sevengill shark	*Heptranchias perlo*
Sand tiger shark	*Carcharias taurus*
Smalltooth sand tiger	*Odontaspis ferox*
Hammerhead shark	*Sphyran spp.*
Tope	*Galeorhinus galeus*
Bull ray	*Pteromylaeus bovinus*

Northridge (1991) reviews several reports of Taiwanese vessels fishing with large driftnets in different areas of the Atlantic Ocean. However, apart from accounts confirming these activities in the Atlantic, no other information is available. There are no reports of the fate of driftnet fisheries in the Atlantic Ocean though all large-scale driftnet fishing was prohibited after 1992 in parallel with the worldwide ban on high seas driftnet fishing.

2.3.1.5 Overview of Driftnet Fisheries

High seas driftnet fisheries have been an important source of elasmobranch by catches. The estimates here suggest that the total elasmobranch by catch could have been between 3 280 000 and 4 310 000 sharks and rays per year during 1989-1991, i.e., of the order of 20 000-38 000t/yr. Total discards of elasmobranchs at sea from driftnet fisheries could have been as high as 30 500t/yr. If all Taiwanese and French catches were kept, discards could have been as low as 20 803t/yr. These results are derived from the estimated totals for the previously described fisheries and thus are highly uncertain. They should be used only as an approximation of the amount of elasmobranchs taken by driftnets worldwide.

Even though these figures are approximate, a clear picture arises from the analysis of information from global high seas driftnet fisheries (Table 2.16). The North Pacific fisheries were the most intensive and therefore the most important in terms of waste of sharks and rays. In particular, the flying squid fishery, with its high catch rates and massive effort, killed more elasmobranchs than any previous high seas driftnet fishery. Of the world by catch of elasmobranchs by driftnets, the North Pacific fisheries accounted for the largest proportion of the total and were also the best studied.

Table 2.16. Summary of estimated bycatch of elasmobranchs in high seas driftnet fisheries.

Fishery	Total catch in tonnes		Total catch in number of individuals	Catch rates (sharks/1000 km nets)
	Lower level	Upper level		
North Pacific Ocean				
salmon(89)	108	1,237	11,492 - 300,000	210 - 5,502
squid(90)	5,905	18,739	2.0 - 2.44 Million	536 - 814
large mesh(90)	1,722	-	151,390	366
South Pacific Ocean(89)	6,381	9,572	56,000 - 841,500*	48 - 181
Indian Ocean(89)	6,108	-	537,000*	-
Atlantic Ocean(91)	430	865	22,000 - 44,000	1,750 - 3,520
Total	20,654	38,243	3,282,882 - 4,313,890	

* from extrapolation of average weight of large mesh fishery

Blue sharks were the most common animal caught in driftnet fisheries because of their abundance in pelagic habitats; in 1989 an estimated 2.2-2.5 million sharks were caught worldwide. Blue sharks may be the elasmobranch most affected by these fisheries but more information is needed to confirm this.

The uncertainty about the estimated catch rates for each fishery highlight the importance of cooperative observer programmes in high seas fisheries: only fisheries with observers provided enough information confidently estimate elasmobranch bycatch and determine the species affected. Hence, the best estimates are those for the North Pacific squid and large-mesh fisheries, the only fisheries which had observers. Greater uncertainty exists in the estimates of catch and mortality rates for the other fisheries.

With the recent closure of large-scale high seas driftnet fisheries, the mortality caused by these fisheries has ceased providing relief to many populations of birds, mammals and other marine fauna. Unfortunately it provides only partial respite for elasmobranchs, particularly sharks, which continue to be caught incidentally in large numbers in other high seas fisheries.

2.3.2 Longline Fisheries

The most important large-scale longline fisheries are those for tunas and billfishes. These fisheries are prosecuted by several countries and occur in all of the oceans. As a consequence of technological innovations such as deep longlines and blast freezing, some of these fisheries supply the most valuable world markets such as that for sushimi. These fisheries target several species and often sharks account for a large part of the bycatches. Sharks are regularly discarded if freezer space, which is limited, is insufficient for the more valuable species. The amount of elasmobranch by catch in these fisheries is unknown and is difficult to assess as most of the international bodies managing these fisheries (i.e. ICCAT, IPTP, SPC, IATTC) do not explicitly include sharks in their statistics or undertake research on elasmobranchs.

2.3.2.1 Atlantic Ocean

Japan, Taiwan (Prov. of China), Korea and Spain have the most important large-scale longline fleets operating in the Atlantic Ocean. Several countries, e.g. Canada, Cuba, USA, Italy, Morocco and Brazil have longline fisheries in their own waters but their efforts are small and in some cases the elasmobranch by catch is utilized and included in official statistics. Most of the information on Atlantic high seas fisheries comes from documents produced by the International Commission for the Conservation of Atlantic Tunas (ICCAT). However, their information is of variable quality; this should be considered when interpreting the results.

Japan

Japanese longliners have fished albacore (*Thunnus elalunga*) and yellowfin tuna (*Thunnus albacares*) in the Atlantic Ocean since the mid 1950's and bigeye tuna (*Thunnus obesus*) since at least 1961. The fleet expanded their range from the western Atlantic equatorial grounds in 1956 to virtually the entire Atlantic by 1970 (Figure 2.35) (Susuki, 1988). Most recently, bigeye tuna made up more than half of the catches and is targeted by deep longlines year-round between 45°N and 45°S. Deep longlines were introduced by the Japanese fishery in 1977 and they also take yellowfin tuna and swordfish (*Xiphias gladius*). Additional effort is directed towards bluefin tuna (*Thunnus thynnus*) in the Mediterranean Sea (ICCAT 1991a).

The number of Japanese longliners in the Atlantic during 1988 and 1989 was reported as 183 and 239 (NRIFSF 1992) and used 68 444 716 and 91 395 915 hooks respectively (ICCAT 1992). Recent data show that the Japanese fleet's effort is increasing in the Atlantic Ocean with 96 651 000 hooks set during 1990 (Uozumi 1993). Japan reported 366 and 500t of "other species" caught in 1988 and 1989 but there is no indication if this includes sharks or other elasmobranchs.

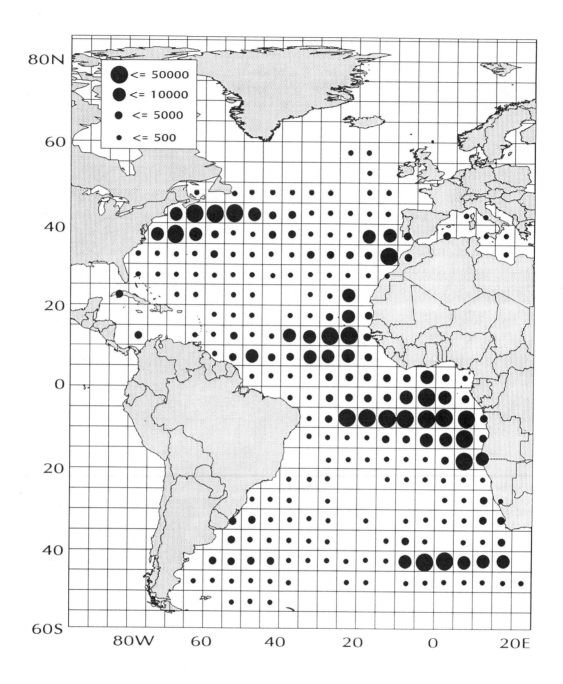

Figure 2.35. Effort distribution of Japanese longline fishery in the Atlantic Ocean in the 1980's. Keys indicate accumulated nominal book numbers in thousands. (Redrawn from Nakano 1993).

Hooking rates of sharks in the different areas of the Atlantic Ocean where the Japanese longliners operate are poorly documented. With one exception, most available information relates only to Japanese longlining activities in the North West Atlantic. Witzell (1985) estimated hooking rates of sharks by Japanese longliners at 1.31 sharks/1000 hooks (107 kg/1000 hooks) for the Gulf of Mexico and 5.98 sharks/1000 hooks (378 kg/1000 hooks) for the US Atlantic Coast. These are minimum estimates as they are based on Japanese logbook information and under-reporting is known to occur (Nakano 1993). Reports from observers in Japanese longliners fishing in the Gulf of Mexico indicate higher hooking rates of 1.74 sharks/1000 hooks (Lopez et al. 1979). Au (1985) documents catch rates of between 1 and 5 sharks/1000 hooks as the most frequently recorded for Japanese longliners in US waters based on observers' data. Au reports about 20 shark species to occur in the by catch.

Hoff and Musick (1990) provide monthly numbers of fish caught for 10 shark groups and numbers of sets made by Japanese longliners in the US EEZ in 1987. They report 8330 sharks, from more than 8 species, taken as by catches in this fishery. Blue sharks comprise about 85% of the total numbers followed by porbeagle and shortfin mako. No indication of sizes or weights is given. Assuming an average of 2206 hooks per set (derived from data of Lopez et al. 1979) the total hook rate is 7.04 sharks/1000 hooks.

Hooking rates reported by Nakano (1993) for sharks in Japanese Atlantic operations range between 1 and 4.5 sharks/1000 hooks with an average of 2.1 sharks/1000 hooks. Nakano lists 11 elasmobranchs (10 sharks and 1 ray) caught during a research cruise in the Atlantic during the 1960's but does not give hooking rates by species. Although Nakano derives separate estimates for the North and South Atlantic, these hooking rates are underestimated because of the common under-reporting of sharks in logbooks. Most skippers do not report sharks catch and some only record sharks of economic value (Nakano 1993).

Information on shark by catches by other longline operations confirms the order of magnitude of hook rates estimated above for the Japanese fishery. Research cruises by the USA in the North Atlantic are documented by Sivasubramaniam (1963) and Brazilian tuna longliners in the Equatorial West Atlantic by Hazin et al. (1990). From Sivasubramaniam (1963) hook rates of 10.35 sharks/1000 hooks can be derived for an area inside 0-80°W and 30-40°N. A smaller area within this had catch rates for blue and oceanic whitetip sharks (*Carcharhinus longimanus*) of 3.32 and 2.3 sharks/1000 hooks respectively. For Brazilian longliners, averages can be calculated from the hook rates for 6 shark groups provided by Hazin et al. for 1° squares off Rio Grande do Norte. The results give an overall rate of 8.66 sharks/1000 hooks, 3.94 for blue sharks, 4.17 for grey sharks (genus *Carcharhinus*), 0.27 for mako sharks, 0.08 for thresher sharks, 0.14 for crocodile sharks (*Pseudocarcharias kamoharai)* and 0.06 for oceanic whitetip sharks. In coastal areas higher hooking rates of up to 41.6 sharks/1000 hooks occurred (Berkeley and Campos 1988).

Extrapolating from these hooking rates for specific areas to the total Atlantic is dangerous as the distribution of sharks is not homogeneous in space and time. Also, two different kinds of gear (regular and deep longline) are used in commercial longlining which have different effects on the catches (Gong et al. 1987, Gong et al. 1989). But, the reported range of hooking rates places bounds on the uncertainty. From the reports listed above there appears general agreement that the hooking rate for the Atlantic Ocean is ranges between 1 and 10 sharks/1000 hooks.

Because of the scarcity of information, hooking rates derived from Hoff and Musick (1990) are used here to estimate total catches of Japanese longliners in the Atlantic. They constitute the most recent data based on Japanese longliners and are well within the overall range of hook rates available. But as the species composition of the shark bycatch changes with location of the fishing grounds, no extrapolation is made to the whole Japanese Atlantic fleet because of the limited areal coverage of Hoff and Musick's data. The figure of Hazin et al. (1990) of 40.91 kg per shark is used to estimate the weight of the catch. The total catch of sharks by Japanese longliners during 1989 in the Atlantic Ocean is estimated as outlined above at 643 427 sharks or 26 322t. The estimates for 1990 are 680 423 sharks or 27 835t. However, some uncertainty is associated with these assessments. The estimates for 1989 could be substantially smaller (14 619t) if calculated using the 30% ratio of sharks to total tuna catches suggested by Taniuchi (1990), or larger (40 149t) if the average weights reported by Witzell (1985) for the South East Atlantic USA area are used. However, the average weight of 40.91

kg/shark seems to be supported by Rodriguez et al. (1988) who found an average weight of 48.9 kg/shark for the bycatches of the Cuban longline fleet operating in the tropical Atlantic during 1973-1985.

The percentage of sharks killed in the Japanese longline fishery is only 7.2% along the Atlantic U.S. coast because of the mandatory release of all bycatches and probably because most of the catches are blue sharks (Witzell 1985). This species, as well as other carcharhinid sharks, survives better when caught by longlines than lamnoid sharks (Sivasubramaniam 1963, Hoff and Musick 1990, Hazin et al. 1990). If this mortality rate is common for the whole Japanese Atlantic fishery, then between 1052 and 2890t of sharks died during their 1989 operations. However, other reports indicate that the U.S. enforced release of all shark by catches in this fishery is not observed for the entire Atlantic (Nakano 1993).

Moreover, the species composition of the by catches changes with latitude and this could alter survival rates. Additional errors in the estimated by catch of elasmobranchs are expected arising from the multiple areas and types of gears used by the Japanese longliners across the Atlantic Ocean. However, as better data on areal, seasonal and gear-specific hooking rates are unavailable it is impossible to obtain better estimates.

The reported catch of elasmobranchs by Japan in the Atlantic Ocean in 1989 is 1540t (Section 2.2). This is close to the lower limit of the range of elasmobranch catch estimated here. However, if the average of the different estimates provided above is taken then at least 15 466t of sharks would have been dumped with most finned prior to discard (Nakano 1993).

Korea

The Korean longlining fleet had 29 vessels operating in the Atlantic Ocean in 1988 and 33 during 1989 (NFRDA 1992). This fleet uses deep longlines which since 1980 have been directed mainly at bigeye tuna. Both the number of vessels and the catches of Korea in the Atlantic have decreased since 1977. These vessels reported an effort of 21 968 198 hooks and a total "others" catch of 944t for 1989 (ICCAT 1992). No information is available on the species composition of the "others" category and no reports of elasmobranch by catches for this particular fishery are known.

An examination of the reported Atlantic fishing grounds of the Korean fleet during 1983-1985 (NFRDA 1988) shows that most of the effort was between 20°N-20°S (Figure 2.36). Thus, it is more appropriate to use the hook rates derived above from Hazin et al. (1990) for the equatorial Atlantic. It is estimated that 190 245 sharks (86 554 blue sharks, 91 607 grey sharks, 5932 mako sharks, 1758 thresher sharks, 3076 crocodile sharks and 1318 oceanic whitetip sharks) or some 7783t were caught during 1989 by Korean longliners in the Atlantic Ocean. This estimate is high compared to the reported 143t of elasmobranchs reported taken in that year by South Korea in the Atlantic Ocean (FAO 1993). Presumably an elasmobranch discard of at least 97% occurred in this fishery. The proportion of sharks released alive and the extent of finning practices in the Korean fishery are unknown.

Taiwan (Prov. of China)

Longliners from Taiwan (Prov. of China) have fished for albacore in the South Atlantic since at least 1967 and in the North Atlantic since at least 1972. More than 80% of their catch is of albacore, Bigeye tuna is the next most common species taken. During 1989, 3 600 000

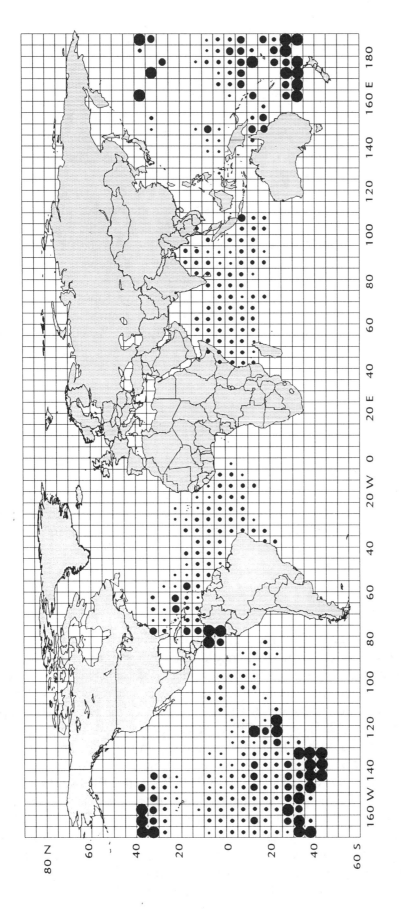

Figure 2.36. Distribution of Korean longline catches, no units given. (Redrawn from NFRDA 1988).

hooks were deployed in the North Atlantic by Taiwan (Prov. of China) compared to 68 700 000 in the South Atlantic (ICCAT 1991b). According to Hsu and Liu (1992) in 1990 this increased to 99 800 000 hooks, 17.4 and 82.4 million in the North and South Atlantic respectively. Of these, 17 500 000 hooks were used by deep longlines fishing bigeye and yellowfin; the remaining 82 200 000 hooks were on longlines fishing for albacore principally in the South Atlantic (Figure 2.37). The Taiwanese catch of sharks was 736t for 1990 and during 1991 the number of vessels operating in the Atlantic fell about 10% though the reported shark bycatch increased to 1486t (Hsu and Liu 1993). Hsu and Liu (1993) note that the variations in the reported by catches of sharks from this fishery are determined by the catch success for target species. When tuna catches are low, vessels keep a larger proportion of the shark by catch.

The reported catch of sharks in this fishery is small for the number of hooks deployed by the Taiwanese longlining fleet. The Taiwanese fish predominantly in the South Atlantic and thus the hooking rates derived from Hazin et al. (1990) are more appropriate. Nevertheless, much of the effort occurs in temperate waters so the amount of by catch can not be separated by species. Under these assumptions, Taiwanese longliners caught an estimated 864 268 sharks in 1990 (equivalent to 35 357t). The actual catch of elasmobranchs by Taiwan (Prov. of China) from the Atlantic Ocean is unknown, thus this analysis can only be approximate. However, it indicates an alarmingly discard of 34 000t of sharks from the fishery! As in the other fisheries discussed here, the number of sharks released alive or discarded dead is difficult to determine with the available information.

Spain

The Spanish longline fishery for swordfish in the Atlantic can be traced from 1973 (Garces and Rey 1984). Fishing grounds for 1988-1991 were centred in the Eastern Atlantic between 55°N and 15°S (Figure 2.38) though some activity was reported in the Mediterranean. Surface longlines are used in the North Atlantic but deep longlines were used in the Southeast Atlantic. The deep longlines consist of baskets of about 1200 m of line between floats having some 33 branch lines 15m long with the deepest hooks between 360 and 470 m (Rey and Muñoz-Chapuli 1991). The Spanish fleet set 35 850 078 hooks in the Atlantic Ocean and 7 683 580 in the Mediterranean during 1989 with increases of 6.75 and 7.3% in 1990 respectively (ICCAT 1991a, 1992).

De Metrio et al. (1984) give catch rates of blue sharks in swordfish longlines in the Mediterranean of 0.014/1000 hooks. However, they do not consider other shark species or discards at sea and the estimates are thus biased downwards. Rey and Alot (1984) give catch success rates for the Spanish swordfish fleet in the western Mediterranean of 6.34 blue sharks, 0.32 shortfin mako sharks, 0.21 smooth hammerhead sharks (*Sphyrna zygaena)* and 0.005 pelagic rays, per 1000 hooks. Mejuto (1985) reports CPUE values of 138.8, 17.5 and 1.1 kg/1000 hooks for blue, shortfin mako and porbeagle sharks respectively in the north and north western grounds of the Spanish Atlantic swordfish fleet based on a sample of 200 trips during 1984. Based on Mejuto's report, this gives catch rates of 13.7, 0.259 and 0.016 sharks/1000 hooks respectively for those species. These catch rates include discards of blue sharks, which Mejuto estimates at 68.4% in weight. Mejuto also found a linear relationship between swordfish catch and discards of blue sharks due to limited storage capacity and low value of blue sharks. He notes that in many cases the shark fins were removed before discarding. More recently, Mejuto and Iglesias (1988) report on exploratory swordfish longlining during 1986 in the Western North Atlantic. Their data gives catch rates of 13.5 and 2.05 sharks/1000 hooks or 168 and 61.7 kg/1000 hooks for blue and shortfin mako sharks respectively.

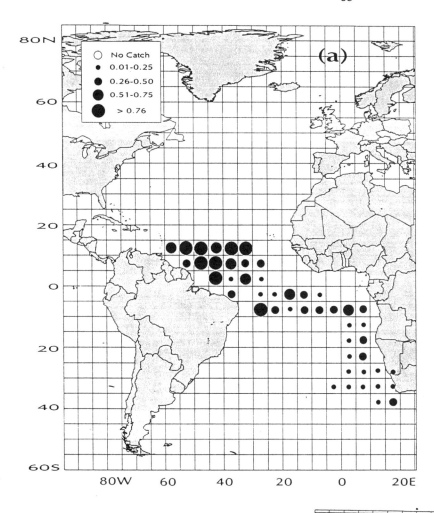

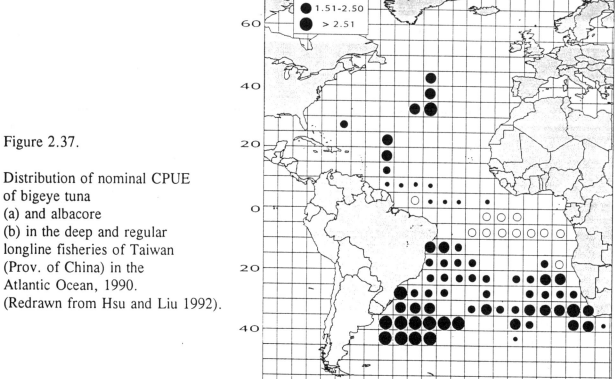

Figure 2.37.

Distribution of nominal CPUE of bigeye tuna (a) and albacore (b) in the deep and regular longline fisheries of Taiwan (Prov. of China) in the Atlantic Ocean, 1990. (Redrawn from Hsu and Liu 1992).

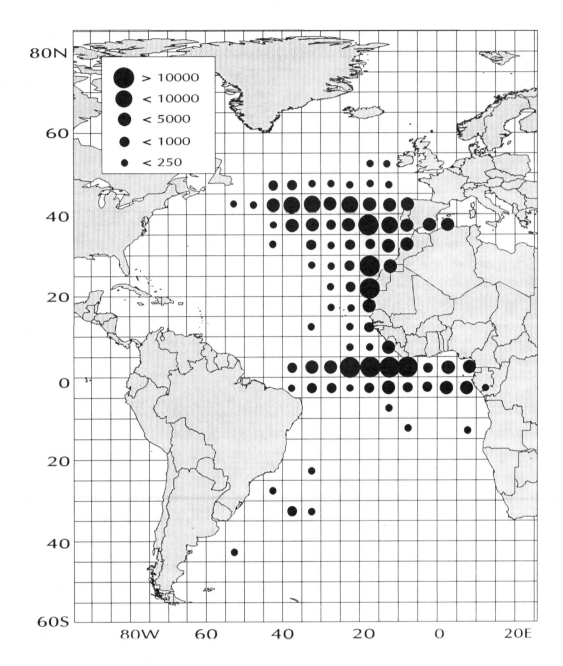

Figure 2.38. Distribution of effort (in thousands of hooks) by the Spanish swordfish longline fishery in the Atlantic Ocean during 1988-1991. (Redrawn from Mejuto et al. 1993).

The elasmobranch bycatch of Spanish longliners includes more than the 3 species mentioned above; Muños-Chapuli (1985b) report 16 species of sharks occurring in the landings of the fleet fishing between Cape Verde Island and the Azores. The blue, the shortfin mako and the smooth hammerhead shark, *Sphyrna zygaena,* were, in order, the most abundant sharks in the catches (Table 2.6, Section 2.2.2).

The limited information from the southern Atlantic fishing grounds of the Spanish swordfish fishery, where deep longlines are used, suggests important changes in the species composition. Rey and Muñoz-Chapuli (1991) report 14 elasmobranch species in the catches of this area from 16 nights fishing of a single commercial longliner. Their data give average shark hook rates per 1000 hooks of 20.6 for night sharks, *Carcharhinus signatus,* 6.3 for silky sharks, 3.4 for bigeye thresher sharks, 2.9 for blue sharks, 2 for devil rays *Mobula sp.,* 1.8 for shortfin mako sharks, 0.3 for common hammerhead sharks and less than 0.3 for *Sphyrna couardi*, *S. mokarran*, *S. zygaena*, *Centrophorus granulosus*, *Galeocerdo cuvieri*, *Isurus paucus* and *Carcharhinus plumbeus*. The overall catch rate of elasmobranchs is estimated at 38.8 fish/1000 hooks which is high compared to those for Spanish swordfish longliners in the North Atlantic. The different areas fished and gears used may cause these discrepancies, but the limited period and few operations observed by Rey and Muñoz-Chapuli could also be a significant source of bias.

The total catch of sharks by the Spanish fishery for 1989 can be estimated using the results of Mejuto (1985). His report, which considers the discards of blue sharks and provides catch rates in weight, is based on a larger time frame and geographic coverage than other reports. It is estimated that with the effort level in 1989, more than 608 000 sharks weighing 6856t were caught by this fishery (5646t in the Atlantic and 1210t in the Mediterranean (Table 2.17). Mejuto also estimates the discard rate to be 68.3% for blue sharks in the Spanish swordfish fleet and finds an inverse relationship between blue shark discards and swordfish catch. The total discard of blue sharks from the Spanish fishery during 1989 was estimated at 4134t. These results

Table 2.17.   Catch rates and estimated total catch of sharks in the Spanish swordfish fishery.

Species	Information from Mejuto (1985)				Estimated total catch 1989			
	Numbers	Weight (t)	Hook rate	CPUE	Mediterranean (7.68 M hooks)		Atlantic (35.8 M hooks)	
	(17.344 M hooks)		(sh/1000 h)	(kg/1000 h)	Numbers	Weight(t)	Numbers	Weight(t)
Prionace glauca *	237,660	2,408	13.703	138.8	105,286	1,067	491,244	4,977
Isurus oxyrinchus	4,488	304	0.259	17.5	1,988	135	9,277	628
Lamna nasus	272	20	0.016	1.1	120	9	562	41
Totals	242,420	2,732	14	158	107,395	1,210	501,083	5,646

* includes estimated discards (68.4%)

should be used with caution as they are based on estimates from only part of the geographical area fished by the Spanish fleet. But they do provide a general indication of the elasmobranch by catches and discards.

2.3.2.2 Indian Ocean

The three principal longline fleets fishing tunas in the Indian Ocean are from Japan, Korea and Taiwan (Prov. of China). They started fishing in 1952, 1963 and 1966 respectively. Indian longliners started fishing for tunas in 1986 but their catches, along with those from the few other fishing countries, was small in comparison (IPTP 1990). Most of the information about longline fisheries in the Indian Ocean is documented in reports of the Indo-Pacific Tuna Development and Management Programme (IPTP).

The Japanese fleet fished tropical areas for yellowfin, albacore and bigeye tunas at the beginning of the fishery but shifted to higher latitudes to target southern bluefin and bigeye tuna during the 1970's, introducing deep longlining in tropical waters at the same time. Judging from data given to the IPTP, Japanese longliners decreased their effort from 106 649 999 hooks in

1986 to 74 861 000 hooks in 1989. The data records of Japanese longliners in the Indian Ocean do not include sharks, so they are not reported by Japan as being caught in the fishery. However, FAO yearbooks cite Japanese catches of 675t of "various elasmobranchs" from the Indian Ocean during 1989. As the only Japanese fishery in those waters is the tuna longline fishery (except for 3 newly introduced purse seiners), the elasmobranch catches reported by FAO, although small, can be attributed to shark bycatches of the longliners.

Taiwanese vessels take the largest catches of albacore but also fish for yellowfin and bigeye tunas primarily using deep longlines in tropical waters (Figure 2.39). A total of 199 vessels participated in the fishery in 1983, decreased to 127 in 1985 and then increasing to 187 in 1988. The total effort, in nominal hooks, during 1988 was 107 million (IPTP 1990). Unpublished data from IPTP show 33 052 sharks with a total weight of 1216t were caught by Taiwan (Prov. of China) in this period using 130 235 742 hooks. For 1989 these values were 188 615 sharks or 7 474 t with an effort of 136 418 296 hooks.

Korean longliners operate primarily in the tropical Indian Ocean (Figure 2.36) targeting bigeye and yellowfin tunas with deep longlines. The number of vessels peaked in 1975 at 185, decreased to 62 in 1985 then increased to 112 in 1988 (IPTP 1990). The most recent data from IPTP, shows they caught 10 851 sharks in 1987 with an effort of 35 748 292 hooks.

The Japanese bycatch of sharks must be estimated as no data are available. Further, the apparent hooking rates derived from the Korean and Taiwanese operations are too low compared with results from the Indian Ocean (see below) and similar fisheries in other oceans (e.g. the Atlantic). The estimated rates were 1.38 sharks/1000 hooks for Taiwan (Prov. of China) in 1989 and 0.3 sharks/1000 hooks for Korea in 1987. The high-grading of catches and discard of sharks in high seas tuna fisheries is common. The results here for these two countries probably reflect considerable under-reporting.

Information on shark by catches in the Indian Ocean longline fisheries is relatively abundant and allows geographical partitioning of the catch in some cases. However, few reports include data on hooking rates by species. The only species composition data is that given by Taniuchi (1990) who reports the percentage of each species in the shark by catches of research tuna longliners from Japan. He shows that 76.6% are blue sharks, 6.6% silky sharks, 6.5% shortfin mako sharks, 3.4% oceanic whitetip sharks and 6.8% unidentified sharks. Sivasubramaniam (1963) provides data on early research operations by Japanese and Taiwanese vessels that indicate bycatches of 10.83 sharks/1000 hooks for the eastern Indian Ocean (E of 60°E). Sivasubramaniam (1964) reports on commercial and research operations for six areas of the Indian Ocean and notes that about 20 species of sharks occur in the bycatches, 11 of these sharks (mainly carcharhinids) are common (Table 2.18). The results of Sivasubramaniam indicate latitudinal changes in species composition of sharks and higher hooking rates for sharks north of the equator. Frequency distributions of hooking rates for sharks are given for 6 areas of the Indian Ocean and show a range of 0-4 to 44.1-49 sharks/1000 hooks. The modal class corresponds to 4.1-8 sharks/1000 hooks. Mimura et al. (1963) provide hooking rates by area and season that average 5.1 sharks/1000 hooks (range 2.6-7.3).

Pillai and Honma (1978) provide monthly catch rates for pelagic sharks in 10°x20° squares of the Japanese fleet in the Indian Ocean that range between 0.1 and 50 sharks/1000 hooks. Varghese (1974; cited by Pillai and Honma, 1978) reports hooking rates as high as 84 sharks/1000 hooks and an average weight of 57 kg/shark in the Lakshadweep Sea. According to Silas and Pillai (1982), hooking rates of sharks in the Indian Ocean vary from year to year and

Table 2.18. Shark species commonly caught by tuna longlining in the Indian Ocean (adapted from Sivasubramaniam, 1964).

Scientific name	Approximate mean weight
Carcharhinus longimanus	30 kg
C. falciformis	60 kg
C. albimarginatus	40 kg
C. melanopterus	35 kg
Prionace glauca	50 kg
Isurus oxyrinchus	75 kg
Lamna ditropis	75 kg
Galeocerdo cuvier	?
Sphyrna spp.	75 kg
Alopias pelagicus	50 kg
A. superciliosus	100 kg

between areas, the highest being between 0.6 and 10 sharks/1000 hooks. They also report that in the Southeast Arabian Sea, sharks were 63.8 and 57.8% of the total catch in number and weight respectively and had an average weight of 30kg. Sivasubramaniam (1987) summarizes data from Fisheries Survey of India tuna research cruises off the south west coast of India during 1983-1986. These results indicate catch rates of 17.6 sharks/1000 hooks. James and Pillai (1987) review additional research cruise result from areas of the Southeast Arabian Sea, Andaman Sea, Western Bay of Bengal and the Equatorial Region south of India. They found the percentage contribution of sharks to the total catch averaged 39.8% (range 30.9-43.7%). They also refer to average catch rates of 16.4 sharks/1000 hooks (range 7.4-29.7) in the Southeast Arabian Sea. James and Jayaprakash (1988) report on two studies of several areas around India. The results indicate catch rates of 8.43 sharks/1000 hooks (range 3.3-14) and a contribution of sharks to the catches of 32.1% (range 19.6-44.8) in one case and catch rates of 7.6 sharks/1000 hooks (range about 1.5-9.5) and contributions of sharks to the catch of 17.4% in the other. Stevens (1992) reports catch rates of 8.3 blue sharks and 3.5 mako sharks per 1000 hooks for a Taiwanese research longliner in south Western Australia.

Strong variations occur in catch rates across the Indian Ocean depending on area and season. Ideally, an estimate of elasmobranch by catches is desired but the aggregated nature of effort statistics for each of the fleets of Japan, Korea and Taiwan (Prov. of China) makes it impossible to apply the appropriate hooking rates for the different regions. However, there seems to be agreement of around 1-10 sharks/1000 hooks as the most common hooking rate. Total catches of sharks in numbers for the whole Indian Ocean can be roughly estimated using a catch rate of 7.96 1000 hooks obtained by averaging the values derived from Sivasubramaniam (1963) and Mimura et al. (1963). These values come from data pertaining to most of the Indian Ocean and also agree with the most common hooking rates reported by different sources. The average weight of sharks taken in the fishery is estimated at 38.2kg derived from the weight and numbers of sharks reported for Taiwanese longliners during 1988 and 1989. The estimated shark by catches for the last available effort levels are: 596 267 sharks or 22 783t for Japan during

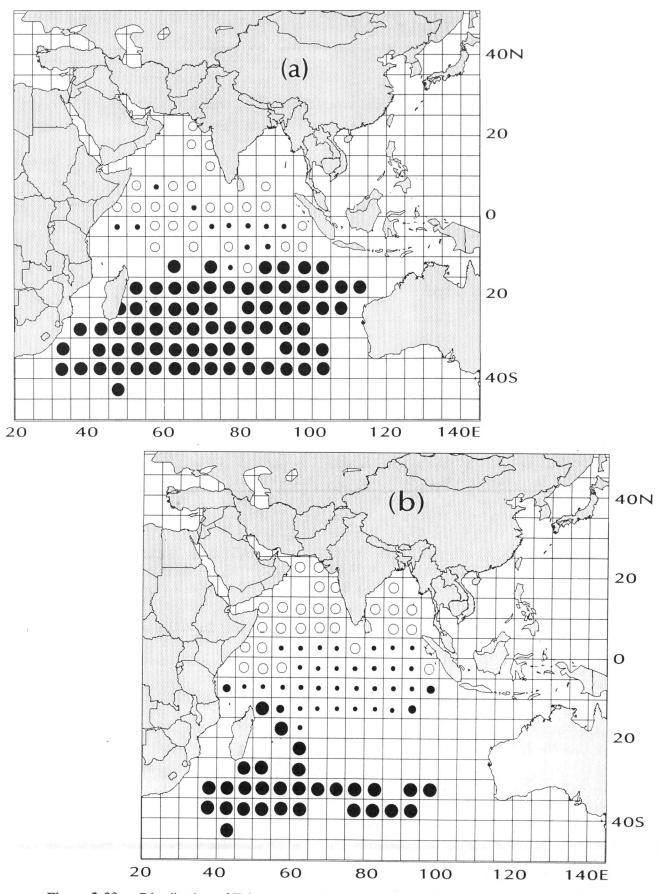

Figure 2.39. Distribution of Taiwanese catch per unit effort of albacore by (a) regular and (b) deep longline fisheries during 1988 in the Indian Ocean. (Redrawn from Hsu and Liu 1990).

1989, 248 735 sharks or 10 879t for Korea during 1987 and 1 086 572 sharks or 41 518t for Taiwan (Prov. of China) in 1989. Thus, a better estimate of the catch of sharks in the Indian Ocean tuna longline high seas fishery is 1 931 574 sharks or 75 180t.

Based on the reported catches of elasmobranchs from each county, the corresponding discards of sharks is estimated at 22 108t by Japan, 9089t by Korea and 34 044t by Taiwan (Prov. of China). The percentages of sharks that survive being hooked and those wasted are unknown but the reports of Sivasubramaniam (1963; 1964) indicate that about 70-80% of the discards of carcharhinid sharks may survive if released alive whereas hammerheads and mako sharks usually die on the line. The rate of finning is also expected to be high. The validity of the estimates are limited by the variability of hooking rates reported for the Indian Ocean and the uncertainty in the effort statistics. Thus they should be used as a first approximation of the amount of elasmobranch by catches and discards in these fisheries.

2.3.2.3 Tropical and South Pacific

Numerous fleets fish for tuna in this area which is home to several small island countries. Most of the longlining is done, in order of importance, by Japanese, South Korean, Taiwanese and Australian vessels. In general, these fisheries are poorly documented making it difficult to ascertain the elasmobranch by catch. Most of the available information for the central Pacific area is that submitted to the South Pacific Commission (SPC) and made available through the Forum Fisheries Agency (FFA)[1]. Australian and New Zealand provide some information about catches in their EEZs but there seems to be no information about areas of the eastern Pacific where neither Australia nor New Zealand have jurisdiction. Further, the cover of the fleets by the FFA data is partial (Lawson 1991). Hence, effort levels for the central and south Pacific area are unknown and are probably larger than those given by the sources used here. The area treated here as Tropical and South Pacific is that south of 20°N.

Japanese fishermen started experimenting with longlines in the western central Pacific as early as the 1920's and 72 vessels were active by 1939. However the peak expansion of this fishery occurred during the late 1960's and covered most of the central and south Pacific (Suzuki 1988, Lawson 1991). At present, at least 406 vessels may operate in the region. The FFA database shows that Japan deploys more than 70% of the total effort in the area, 31 143 fishing days in 1989. The South Korean longline fleet started fishing in 1958 and is reported to have 124 vessels active in the area. According to NFRDA (1988), their longliners fish largely for tunas in the South Pacific (Figure 2.36). The South Korean effort in the FFA zone was 6312 fishing days in 1989. Activities of the Taiwanese fleet are less well documented and not even approximate numbers of active vessels in the region are available. They operate in the waters north of Papua New Guinea and around Fiji and American Samoa (Lawson 1991). According to FFA data, the Taiwanese fleet effort was 4163 fishing days in 1989. The Australian longline fisheries for tuna date back to the 1960's. It expanded in the 1980's to more than 91 vessels by 1989, with a total of 2244 fishing days. In addition to these fleets, a few vessels from China, Fiji and Tonga also operate but in 1989 their effort only accounted for 558 fishing days. The geographical distribution of total longline effort during 1990 available to the SPC is shown in Figure 2.40. Most of the fishing effort occurs between 15°N and 15°S.

---

[1] P. Tauriki, FFA, P.O. Box 629, Honiara, Solomon Islands, pers. comm. June 1992)

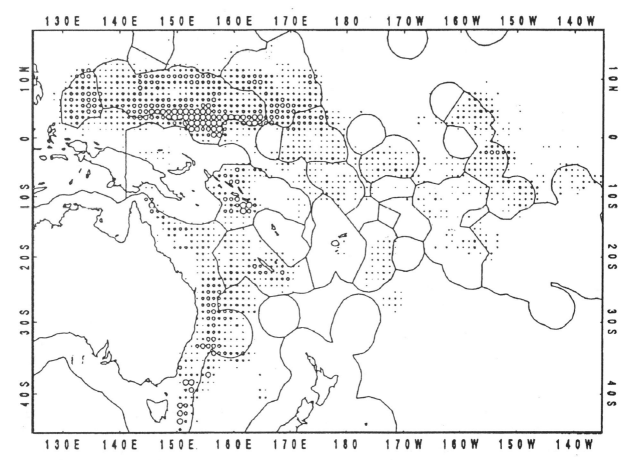

Figure 2.40. Distribution of longline effort in the SPC area during 1990, units not given. (Taken from Lawson 1991).

The reported catch of sharks for 1989 in this area was 426t; 375t by Taiwan (Prov. of China), 35t by South Korea and 12t by the Japan. Although numbers of hooks deployed by country were not available, the total for all longliners was 98 832 500 during 1989. The number of hooks per country can be estimated using the reported fishing days of each fleet. The corresponding estimated catch rates in kg/1000 hooks are 0.167 for Japan, 2.5 for South Korea and 40.5 for Taiwan (Prov. of China). This is equal to an overall catch rate of 4.31kg of shark per 1000 hooks. Such minuscule catch rates, equivalent to less than 0.5 sharks/1000 hooks, are almost certainly a result of under-reporting, presumably due to discarding. This is evident in the comparison of the estimated catch rates for each of the countries.

Saika and Yoshimura (1985) plot hooking rates for the most common sharks taken by Japanese research longliners in the western equatorial Pacific. These are approximately 0-14/1000 hooks for oceanic whitetip and for silky sharks, 0-16/1000 hooks for blue sharks and 0-2/1000 hooks for shortfin mako sharks. An overall rate of 20.45 sharks/1000 hooks can be obtained for waters below 22°N from the report of Strasburg (1958) on research and commercial cruises in the eastern equatorial Pacific. This can be further split into 4.14 for blue sharks, 5.46 for oceanic whitetip sharks, 10.07 for silky sharks and 0.78 for unidentified sharks, per 1000 hooks.

Stevens (1992) gives by catch data for blue and mako sharks by longliners fishing off Tasmania from observers onboard Japanese vessels targeting mainly southern bluefin tuna (*Thunnus maccoyii*). These data show catch rates of 10.4 for blue sharks and 0.5 for mako sharks per 1000 hooks. Stevens estimates that 1594 mako and 34 000 blue sharks weighing 24 and 275t respectively, are caught each fishing season in this fishery. Hooking rates for other species are not available but Stevens mentions that thresher, porbeagle, school (*Galeorhinus galeus*), black (*Dalatias licha*), crocodile (*Pseudocarcharias kamoharai*), hammerhead, velvet dogfish (*Zameus squamulosus*) and grey (*Carcharhinus*) sharks are also present in the by catches of Japanese longliners in the Australian Fishery Zone. He also provides data for the by catches of blue sharks in New Zealand waters where the Northern New Zealand Japanese and Korean fisheries had catch rates of 4.8 and 1.3 blue sharks/1000 hooks respectively and the southern New Zealand Japanese fishery, catches 5.4 blue sharks/1000 hooks. Stevens notes the under-reporting of shark by catches in Japanese logbooks and reports that fins are removed from the sharks before being discarded. If so, the mortality in this fishery would equal the total estimated by catch.

Ross and Bailey (1986) provide hooking rates for mako sharks in the northern New Zealand Korean and Japanese fisheries for albacore and for the southern New Zealand Japanese fishery for southern bluefin tuna. Averages are 0.43 and 0.34 sharks/1000 hooks for the northern and southern fisheries respectively. Based on their data, the estimated catch of mako sharks is 334t processed weight. As about 50% of a shark's weight is lost during processing, the estimated live weight of the mako shark by catch is 668t. Ross and Bailey provide no further information and this estimate may only represent the reported catch and not discards.

The total by catch of sharks in the SPC zone can be estimated using figures estimated from Strasburg (1958) and a conservative estimate of 20 kg/shark to calculate the total weight. Even though this catch rate might be too high, the distribution of effort in these fisheries (see Figure 2.40) justifies the use of the hooking rates from the Equatorial Pacific. The results (Table 2.19) indicate that approximately 2 021 711 sharks, or 40 434t, were caught in 1989 and almost 50% of these were silky sharks. Japan takes the majority of the elasmobranch catch and also has the highest discard rate. Total discards are estimated at 40 000t.

Table 2.19. Estimated bycatch of sharks in tuna longline fisheries of the Central and South Pacific (SPC zone), based on the results of Strasburg (1958).

Species	Strasburg's data		Estimated Catch in 1989					
	Numbers caught (216,172 hooks)	Hook rate (#/1000 hooks)	Total numbers	weight (t)	Japan weight (t)	S. Korea weight (t)	Taiwan weight (t)	Australia weight (t)
Carcharhinus falciformis	2,176	10.07	994,854	19,897	13,950	2,827	1,865	1,005
Carcharhinus longimanus	1,181	5.46	539,946	10,799	7,571	1,535	1,012	546
Prionace glauca	896	4.14	409,646	8,193	5,744	1,164	768	414
Various sharks	169	0.78	77,266	1,545	1,083	220	145	78
Totals	4,422	20.46	2,021,711	40,434	28,349	5,746	3,789	2,043

Shark by catches for the entire tropical and South Pacific might be higher. Judging from the size of the statistical area covered by the SPC (Figure 2.34) and the maps of CPUE of the South Korean longline fleet for 1983-1985 (Figure 2.36) and considering the partial coverage of the SPC area by FFA statistics (SPC 1991), it is estimated that the South Korean fleet deployed twice as many hooks in the whole central and South Pacific as those reported by the FFA; similarly for the Japanese and Taiwanese fleets. In this case, the estimated catch of sharks in the central and south Pacific outside the SPC zone is 1 097 288 sharks or 21 946t; 16 422t by Japan, 3328t by South Korea and 2196t by Taiwan (Prov. of China). These figures assume an extra effort of 92 598 173 hooks (1989) and a total catch rate of 11.85 sharks/1000 hooks. This rate considers the possible effort less higher in latitude areas and is calculated by averaging the hooking rates obtained from Strasburg (1958) for the equatorial zone, those of Stevens (1992) for Tasmanian waters and those of Ross and Bailey (1986) and Stevens (1992) for New Zealand waters. The same weight of 20 kg/shark is used. Thus, it is estimated that 62 380t of sharks were caught as by catch of longline fisheries in the whole central and south Pacific in 1989. According to FAO statistics, the total reported catch of elasmobranchs from the West Central, South Western and South Eastern Pacific of Japan, Taiwan (Prov. of China) and Korea was only 4409t for 1989. These figures suggest that some 58 000t of sharks may be discarded. These estimates could be less uncertain than those calculated in previous sections for other high seas longline fisheries. Because of the limited information available about the real effort levels of each fleet and the hooking rates in the South Pacific.

2.3.2.4 North Pacific

This is another area where longline fisheries activities are poorly documented. CPUEs of Korean longliners published by NFRDA confirm that there was some effort by this fleet in the central north Pacific during 1983-1985 (Figure 2.36). Figures from Suzuki (1988) show that the Japanese longline fleet operated in the north Pacific. Though, Taiwan (Prov. of China) does not have a high seas longline fishery in this area (Nakano and Watanabe 1992). No statistics are available, at least in English, on the amount of effort deployed by longliners in the North Pacific.

Nakano and Watanabe (1992) estimate the longline effort of the Korean fleet at 14-19 million hooks/yr for 1982-1988. Using this estimate and statistics from the Fishery Agency of Japan they estimate a total effort of 258 422 780 hooks deployed during 1988 in the entire North Pacific by Japan and Korea. Their estimate of 3 274 609 blue sharks caught by longline fisheries in the North Pacific during 1988 is based on latitudinal stratification of effort and hooking rates. Because of the geographical coverage considered in the previous section for the Tropical and South Pacific, only waters north of 20°N are considered here as "North Pacific." From Nakano and Watanabe's data is estimated a total effort of 105 885 418 hooks and a by catch of 2 964 500 blue sharks for the North Pacific during 1988.

Data in reports of Strasburg (1958) gives an overall hooking rate of 18.45 for blue sharks, 0.07 for oceanic whitetip sharks and 0.84 for unidentified sharks (total of 19.36 sharks/1000 hooks) for the eastern Pacific north of 22°N. Data given by Sivasubramaniam (1963) indicates hooking rates of 6.79 for blue sharks and 0.35 for oceanic whitetip sharks/1000 hooks for two combined areas of the Pacific north of 20°N. Saika and Yoshimura (1985) present data on shark by catches of Japanese research cruises from 1949-1979 in the Western Pacific. Their maps of hooking rates indicate values of approximately 0-3 oceanic whitetip, 0-0.5 silky, 0-2 shortfin mako and 0-30 blue sharks per 1000 hooks for the region north of 20°N. Catch values plotted for blue sharks appear to be around 10 sharks/1000 hooks whereas the other species probably average to less than 1 shark/1000 hooks. Nakano et al. (1985) provide numbers of blue sharks caught and number of stations sampled for longline cruises during 1978-1982 in the western north Pacific. The longlines utilized had between 1500-1800 hooks. Assuming a mean of 1650 hooks per station, then hooking rates averaged 17.62 blue sharks/1000 hooks which is similar to the estimate derived from Strasburg's data.

The estimated by catch of sharks by tuna longlines in the North Pacific is comparatively high. Based on the hooking rates derived from Strasburg (1958) and the estimated effort from Nakano and Watanabe (1992), a total of 2 050 136 sharks are estimated to have been caught during 1988 in the North Pacific. Roughly 1 950 000 of these would be blue sharks, 7250 oceanic whitetip sharks and about 90 000 sharks (Table 2.20). These estimates for blue sharks taken in the same area are conservative compared to those of Nakano and Watanabe. Assuming an average weight of 20 kg/shark regardless of species, the estimated total by catch is 41 000t. National catch is difficult to estimate since it is impossible to separate, the estimates of effort of Nakano and Watanabe. A crude estimate based on proportions indicates that 7.35% of the catches could be South Korean and the rest Japanese.

Table 2.20. Estimated bycatch of sharks in the North Pacific by the longline fleets of Japan and Korea based on the results of Strasburg (1958)

Species	Strasburg's data*		Estimated Catch in 1988	
	Numbers caught (87,595 hooks)	Hook rate (sharks/1000 hooks)	Total numbers	weight (t) **
Prionace glauca	1,616	18.45	1,953,432	39,069
Carcharhinus longimanus	6	0.07	7,253	145
Various sharks	74	0.84	89,452	1,789
Totals	1,696	19.36	2,050,136	41,003

* for cruises north of 21 N

** assuming 20 kg/shark

There is no information on discards of sharks from these fisheries or the amounts released alive. Given the manner of partitioning FAO statistical areas in the Pacific it is difficult to assign to area, catches of elasmobranchs reported by Japan and Korea. Even considering the total reported "various elasmobranchs" catch of 15 537t for Japan and 2927t for Korea, which correspond to a much larger FAO areas 61, 67 and 77 of the Pacific Ocean, the estimated discard would be of about 22 000t.

2.3.2.5 Overview of Longline Fisheries

High seas longline fisheries for tunas and billfishes are a large source of by catch and discards of elasmobranchs. Despite the uncertainty of the different estimates, it is evident that the amount of effort exerted by longline fleets (worldwide total of about 750 million hooks) is the main cause of the high by catch. The best estimates given in Table 2.21. The total high seas catch by longlines worldwide is estimated at 8.3 million fishes, equivalent to 232 425t! This almost a third of the world catch of elasmobranchs reported by FAO in 1991.

The by catch of blue sharks from longline fisheries is large. Although a species breakdown was not always possible, an approximation can be done for areas where only total shark by catch was estimated if a conservative estimate of 40% of the total is used for blue sharks. Adding this estimate to the numbers of blue sharks caught where a species breakdown is done, gives a total of 4 075 162 blue sharks caught incidentally by world high seas longline fisheries.

Table 2.21.   Selected estimates of shark bycatches in high seas longline fisheries.

Area	Number of individuals	Total catch in tonnes
Atlantic Ocean	2,305,940	76,318
Indian Ocean	1,931,574	75,180
South/Central Pacific Ocean	1,996,350	39,927
North Pacific (above 20N)	2,050,135	41,000
Total	8,283,999	232,425

The relative importance of shark by catches, in number of fishes is almost equally distributed in the longline fisheries of the world. The fisheries of the Atlantic, Indian, Tropical and South Pacific and North Pacific Oceans each account for about 2 million elasmobranchs. However, the total weight of by catch in the Atlantic and Indian Oceans is estimated to be almost double that for the whole Pacific Ocean (Table 2.21). Because of the different mean weights used in the calculations and does not necessarily represent a real difference in weight of the catches. Specifically, the mean weight of 20kg/shark used for Pacific fisheries is conservative.

The amount of discarded sharks and survival rate of released sharks are also uncertain. The accumulated estimates of discards from the longline fisheries treated above amount to 204 347t. It is unknown what proportion of these discards survive but some reports indicate it could be as high as 66% (Berkeley and Campos 1988). Nevertheless, numerous accounts of finning exist in the literature (e.g., Mejuto 1985, Nakano 1993) and given the rise in shark fin prices in the late 1980's it would be naive to think that released sharks are not finned. Further research is needed to determine the mortality of sharks due to longline fisheries.

The present estimates seems to be in agreement with previous assessments. As a reference, Taniuchi (1990) estimates a total shark catch from Japanese longliners of 90 000t using an estimate of the ratio of shark-catch/target-species catch for the tuna and billfish longline fishery. The world elasmobranch by catch estimated here for Japanese longliners is 115 441t. But there is a good degree of uncertainty introduced by the low quality of the baseline information that is available. For example, the hooking rates used here ranges between 7.04-20.45 sharks/1000 hooks whereas Taniuchi (1990) plots rates for Japanese research longliners that range between 2.7 and 8 sharks/1000 hooks. Only reliable regional effort figures and updated hooking rates representative of each region will provide better estimates of the by catches.

In contrast to driftnet fisheries, there are no observer programmes for high seas longline fisheries in the world. This results in much the uncertainty surrounding the estimates of non-target species caught in longline fisheries. Most of the international tuna organizations and the governments of longline fishing nations requiring logbook reports from longline fleets still do not require, or enforce, reporting of by catches of sharks or other elasmobranchs though some organizations are starting to change (ICCAT 1993b, Nakano 1993). This will reduce uncertainty about the levels of by catches and discards in the future. Considering the common underreporting of elasmobranchs in longliner logbooks (Stevens 1992, Nakano 1993), observer programmes are undoubtedly the best way to provide this crucial information.

2.3.3 Purse Seine Fisheries

Most the large-scale purse-seine fisheries for tuna occur in tropical waters where the relatively shallow schooling behaviour of some tunas makes them vulnerable to this type of gear. The main species targeted by this method of fishing are yellowfin (*Thunnus albacares*) and skipjack (*Katsuwonus pelamis*) although other species of tuna, other fish and marine mammals) commonly associated with the schools of tuna, are also frequently caught. Major tuna purse seine fisheries are fairly localized activities. They are centred in four areas: the Eastern Tropical Pacific (ETP), Mexico to the north of South America; the Western Central Pacific (WCP), from the Philippines and Papua-New Guinea to Polynesia; the western Indian Ocean, around the Seychelles and the eastern tropical Atlantic around the Gulf of Guinea (Figure 2.41). Some tuna purse seining also occurs off Venezuela in the western Atlantic Ocean.

The ETP fishery began during the 1950's and expanded in the 1960's and 1970's. In the early 1980's it suffered a temporary decline and today about 280 000t of yellowfin tuna are caught by purse seiners in this region (Sakagawa and Kleiber 1992). The fleet used to be dominated by US vessels but since the early 1980's many of these switched to the WCP fishery and now Mexican vessels are dominant. Tuna purse seining was started in the WCP by Japanese and USA vessels in the 1970's. In contrast to the ETP, the effort here is largely directed towards skipjack although yellowfin are also caught in large amounts. The Japanese fleet mainly fishes log-associated schools while US boats concentrate on free-swimming schools (Sakagawa and Kleiber 1992). Korean and Taiwanese purse seiners joined the fishery in the late 1970's (Suzuki 1988). A smaller number of Australian, Indonesian, Philippine, Marshall Island, New Zealand, Solomon Island and the former USSR vessels also participated. The total purse seiner tuna catch in the WCP during 1989 was 576 204t; at least 73% was skipjack (Lawson 1991).

The purse seine fishery was initiated in the western Indian Ocean (WIO), by a Mauritius-Japan purse seiner in 1979 followed by French vessels in 1980. By 1984 the French fleet together with part of the Spanish fleet moved from the Atlantic to the WIO . During 1989, France, Spain, Panama, Japan, Mauritius, U.S.S.R. and Cayman Island had 49 purse seiners operating in this fishery. The first two countries dominate the fleet. The total catch in the WIO for 1989 was 220 000t, mainly yellowfin and skipjack but also some bigeye (IPTP 1990).

Purse seine fishing for tunas in the tropical Atlantic was initiated by the French in the early 1960's in the coastal waters of the Gulf of Guinea. African coastal states, Spain and US fleets joined later. The fishery expanded to offshore areas at the end of the 1970's and it currently accounts for more than 80% of the Atlantic yellowfin tuna catch (Suzuki 1988). The majority of the catches are now taken by Spanish and French-Ivorian-Senegalese-Moroccan (FISM) fleets with small amounts by Venezuelan, U.S.S.R. and Japanese boats. Yellowfin and skipjack are the main species taken with minor amounts of bigeye tuna taken incidentally. A total of 167 800t of tunas was caught by purse seiners in the tropical Atlantic during 1989, at least 90% of this from the eastern Atlantic (ICCAT 1991a, 1991b, 1992).

Information on elasmobranch by catches in purse seine tuna fisheries is scarce and poorly documented. Even though the presence of sharks in the purse seine catches is documented, at least since the mid-1960's, it has received little attention in the literature. Bane (1966) reports several large silky, as well as other, sharks and devil rays in a set off Gabon in 1961. Bane also mentions that *C. limbatus, C. plumbeus* and *Rhizoprionodon acutus* are associated with tuna schools in the area. Yoshimura and Kawasaki (1985) report 183 silky sharks caught by purse seine in the WCP and length frequency histograms indicate that most silky sharks were between 60 and 170cm TL with the mode at 110-130cm TL. In the Indian Ocean, LaBlache and Karpinski (1988) based on observer's data, give rates of 6% of the total catch for purse seiners

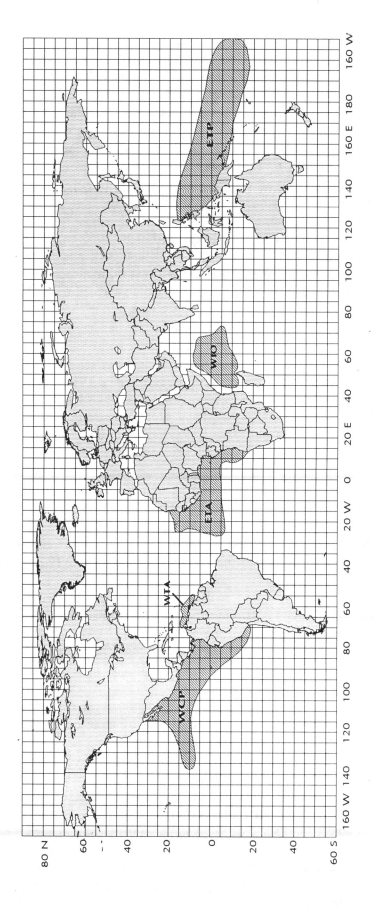

Figure 2.41. Major areas of Tuna purse-seine fisheries in the world.

that had shark bycatches. They consider various teleosts, including undersized and damaged tuna, to comprise the by catch. Oceanic whitetip sharks were the second major by catch (12%).

The most detailed account of sharks associated with tuna schools for the ETP is that of Au (1991). He notes that sharks associate with yellowfin, perhaps as opportunistic predators or scavengers. The percentage of sharks associated with yellowfin measured as percentage of sets having sharks is 40% for log-associated tuna schools, 6-21% for free swimming schools and 13% for dolphin-associated schools. Apparently, these associations are limited by the swimming speed of sharks. Silky sharks were the most common elasmobranch in the by catches with up to 500 individuals caught per set. Various other carcharhinids, oceanic whitetip, sphyrnid, alopid, lamnid, blue and whale sharks were also caught together with various batoids and mobulids. Au's report does not provide any useful measure of the numbers of sharks caught by purse seine fisheries (i.e. catch of sharks per unit of effort, or the relation between tuna catch and elasmobranch catch. Although he lists average numbers of sharks per set by species, these values are based on purse seine sets that caught the pertinent species. Without reference to the total numbers or weights of sharks in the full sample, his results are of limited use for estimating shark by catch rates although they give the species composition of the elasmobranch catch.

The total by catch of elasmobranchs in purse seine fisheries can be estimated using the information on shark and tuna catch provided by Lablache and Karpinski (1988). Their data permits an estimate of shark catch of to 0.51 % of the tuna kept by purse seiners. Using this proportion and the reported tuna catches listed above, the estimated total catch of sharks in purse seine fisheries during 1989 is of 6345t: 856t in the tropical Atlantic, 1122t in the fisheries of the Western Indian Ocean, 2939t in the Western Central Pacific fisheries and 1428t in the Eastern Tropical Pacific. These estimates assume that the amount of sharks caught is proportional to tuna catch. Purse seining is an active fishing method that takes advantage of the schooling behaviour of fish. Sharks gather around tuna schools, especially certain types of schools such as those associated with log (Au 1991). Unlike passive gears, shark catches in purse seine fisheries are not possible without tuna catches. Thus it is appropriate to relate the shark catch to the tuna catch rather than to a measure of effort, e.g. days at sea, where, for passive gear, competition occurs for hooks or space in the gillnet. The main weakness of the present estimates are the calculations of shark catch rates in tuna purse seine operations and the extrapolation from Western Indian Ocean data to other geographical areas.

There are no records of the condition of the elasmobranchs caught in tuna purse seine operations, but it is likely that they die either by suffocation or crushing if they do not bite their way out of the nets. Bane (1966) reports that shark catches were sold in the Gulf of Guinea but this seems to be an exception for an experimental fishing campaign. Most shark catches in tuna purse seine fisheries are probably discarded though this has not been confirmed.

2.3.4 Other miscellaneous fisheries

Other fisheries take elasmobranchs incidentally and although they are either of minor scale or their bycatches are insignificant, it is worth mentioning some of these which might, with time, affect particular elasmobranch stocks. Pole and line fisheries for tunas take some shark bycatches while fishing tuna schools (Anderson and Teshima 1990). Almost nothing is known about the catch rates. Bane (1966) mentions sharks taken by "tuna clippers...at the surface on live bait", which suggests pole and line fishing: 131 sharks were taken at 6 stations by this method. It is possible, due to the global scale of pole and line fisheries for tunas, that their bycatch of sharks could be significant, perhaps in the order of that from purse seiners. Alternatively, pole and line gear may avoid the capture of sharks and survival of discards could be high. These hypotheses could be verified by interviewing skippers from this type of fishery.

The orange roughy (*Hoplostethus atlanticus*) fishery of New Zealand takes deep water squaloid sharks and other elasmobranchs. Although no estimates of catch rates are available, some information exists from research vessels. At least 21 elasmobranchs (11 selachians, 4 batoids and 6 holochephalans) have been identified in deep water trawl surveys around New Zealand (Robertson et al. 1984). There are 8 squaloid sharks of potentially commercial importance, of which *Deania calcea* is the most abundant in the North Island, *Etmopterus baxteri* in the South Island and *Centroscymnus spp.* in the central areas. Surveys carried out in the North Island show that *Deania calcea* constitutes a larger part of the total catches than either the orange roughy or the hoki (*Macruronus novaezelandiae*), currently the most important commercial species (Clark and King 1989). Although catch rates in commercial trawling should be smaller than those of research cruises due to more targeted fishing, it is possible that the by catches of elasmobranchs constitute between 10 and 50% of the orange roughy catches. According to FAO statistics, orange roughy catches in New Zealand waters were of around 44 000t/yr in 1984-1989. The total bycatch of squaloid sharks could therefore be between 4400 and 22 000t/yr in this fishery. King and Clark (1987) estimate the MSY for these sharks as 2250t/yr. Evidently, the current catches far exceed the MSY. Most of the catch is discarded as there is no market though small quantities are used for fishmeal and liver oil extraction. Given the depth at which these sharks are caught (600-1200 m) and the gear employed, all will be dead when returned to the sea.

The impact of this level of bycatch on the local stocks sharks is unknown but it must be highly damaging and likely to lead to unsustainable exploitation. But this is difficult to verify as little information exists about the biology and population dynamics of these species. More research is needed on the levels of by catch, survival of discards and the deep sea shark populations themselves.

2.3.5 Overview

The amount of elasmobranchs caught and discarded in high seas fisheries worldwide is uncertain as neither process is adequately documented. Discard and survival rates are unknown. There are large uncertainties about the catch rates for each region and sometimes also about effort levels. Qualitative and quantitative variations in the elasmobranch bycatches within each ocean due to areal and seasonal changes in availability of the different species should be expected. Present results indicate that a large amount of elasmobranchs are caught incidentally in the high seas fisheries of the world. The estimated annual elasmobranch by catch at the end of the 1980's is around 260 000 and 300 000t or 11.6-12.7 million fish. Most of the catch are sharks, predominantly blue sharks.

Longline fisheries are the most important source of shark kills in the high seas, mainly because of the magnitude of their effort. They contribute about 80% of the estimated total elasmobranch by catch in weight and about 70% in numbers of fish. There is great uncertainty around the estimates for this type of fisheries, but the figures are based on the best available information and seem to compare well with the few reference points available (Section 2.3.2.5). The former high seas driftnet fisheries ranked second in their contribution to the elasmobranch by catches. Since their activities were stopped at the end of 1992 they are now one less worrisome in terms of sea-life conservation. It is conceivable that this effort has been redirected to fisheries which might still affect elasmobranchs and the other species previously affected by their gillnetting activities.

Discards from high seas fisheries also are high. Up to 230 000-240 000t of elasmobranchs are discarded annually by various high seas fisheries. Most discards, certainly those caught by the driftnet, purse seine and orange roughy fisheries, probably die. For longline fisheries, survival depends on whether fishermen release sharks quickly and unharmed, though finning will prevent survival. The little information available on purse seine and pole-and-line tuna fisheries and the deep trawl fisheries for orange roughy make it very difficult to assess the importance of their by catches of sharks and rays. Another source of by catch and waste of sharks and rays is the incidental catch by bottom trawling vessels fishing for shrimps and fishes on continental shelves. The assessment of the impact of these fisheries is difficult because of the difficulty in gathering information about them. These fisheries have high local impacts on populations especially in the case of rays. Some of the elasmobranchs caught are landed and reported under official statistics of the fishing country but a large proportion is discarded and never recorded.

2.3.5.1 Species of Elasmobranchs under Pressure from High seas Fisheries.

Blue sharks are the most common elasmobranch caught incidentally in high seas fisheries; an estimated 6.2-6.5 million blue sharks are taken annually. Although this is apparently the first estimate of total catches for blue sharks in high seas fisheries, some partial estimates are available for comparison, e.g. Stevens (1992) estimates that the Japanese longline fisheries annually take a total of 433 447 blue sharks. His figure is small compared with that estimated here. However, he uses a hooking rate of only 1 shark/1000 hooks. Nakano and Watanabe (1992) estimate that the high seas fisheries of the North Pacific Ocean caught 5 million blue sharks during 1988, an estimate higher than was derived here.

Lack of knowledge prevents an assessment of the impact of the removal of 6 million blue sharks annually on high seas ecosystems or on the blue shark populations. Little is known about the size of the stocks of blue sharks in the world and the biology of most populations is poorly understood. Nakano and Watanabe (1992) provide the only assessment known of the impact of high seas fisheries on blue shark stocks. By estimating bycatches and using cohort analysis, they believe that the catch levels during the late 1980's did not have a significant impact on the populations of the North Pacific. However, Wetherall and Seki (1992) and Anonymous (1992) consider that appropriate information is lacking for an assessment of this kind. Regardless, research is needed to assess the real by catch levels in each fishery and their impacts on the different populations.

Silky sharks are probably the second most commonly caught species, especially in longline and purse seine fisheries. As for blue sharks, little information is available to assess the impacts of removals. Silky sharks have slower growth, later sexual maturation and are less fecund than blue sharks (Pratt and Casey 1990) and hence will be less resilient to exploitation. Local stocks of *Deania calcea, Etmopterus baxteri* and *Centroscymnus spp.* in New Zealand could also be threatened by large-scale fisheries.

3. DISCUSSION

3.1 Current Situation of Elasmobranch Fisheries.

Fisheries for sharks and rays are common throughout the world and differ in both the species taken and in the type of gears and vessels used. This diversity contributes to the difficulty in studying the fisheries and to the problems of collecting accurate data on yields and fishing effort. This is evident from the scarcity of information about most of the cases reviewed here. Few countries have sufficient information on their shark and ray fisheries for assessment purposes. Statistics for elasmobranchs around the world need to be improved: major species and species groups in the catch should be recorded and the elasmobranch bycatch from bottom trawl and high seas large-scale fisheries should be reported. This is best done through observer programmes on high seas fishing vessels and the inclusion of sharks in research programmes and statistical requirements (logbooks) of major international tuna programmes, e.g. ICCAT, IPTP, IATTC and SPC. Much data compilation and reviewing must be done on a country and regional basis to enable appraisal of exploitation levels and to make assessments of the status of elasmobranch stocks. This will require coordinated efforts of fisheries managers, shark specialists and volunteers in each country and region.

Another important characteristic is the predominantly incidental nature of the elasmobranch catch. The number of fisheries which primarily target sharks or rays is few. The majority of fisheries taking sharks and rays, are targeted at other species which makes assessment, and especially management, difficult to achieve. Few managers will constrain economically or socially important fisheries to manage elasmobranchs stocks.

The increasing global trend in reported shark and ray catches suggests that yields will be continue to rise as there is no evidence of decline in production. This is misleading if interpreted uncritically as there is a change in the types of fisheries and species exploited; while some fisheries for elasmobranchs fall, others increase. This indicates that exploitation levels are not sustainable in many cases. Almost 30% of the major fishing countries analyzed in section 2.1.2 show a falling trend on catches. Reasons for an apparent increase in catches could be increases in reporting and more landings of by catches from other fisheries.

Although the analyses of trends in yield in each FAO fishing area (section 2.1) suggest that an expansion of the catches could be achieved from some stocks, in the Northern Indian Ocean, the North Sea and North East Atlantic stocks are probably overexploited. These analyses are approximate and a better index of relative production could be developed to provide a better assessment of the possibilities for increased elasmobranch exploitation. A simple improvement would be to incorporate in the index of relative production the area of continental shelf included in each Major Fishing Area to weigh the production of sharks and rays in a similar way in that which as the surface of sea of each area is used here.

The likelihood that fisheries for elasmobranchs will be sustainably exploited in the near future is not promising as general lack of management and research directed towards these resources is evident in most cases. Only three of 26 major elasmobranch-fishing countries (Australia, USA and New Zealand) are known to have management and research programmes for their shark or ray fisheries. Not one them play a leading role in worldwide elasmobranch production. Moreover, those few countries with fisheries information have apparent problems of over exploitation for some elasmobranch stocks (e.g. shark fisheries in souther Brazil, on both coasts of the USA and in southern Australia). Many of the countries with major elasmobranch fisheries have very limited or non-existent research programmes and probably no management for these resources. If this situation continues stocks will eventually be driven to such low population levels that fishing will probably cease for a very long time. A particular case is Indonesia, where catches have grown quickly in the last 20 years and will probably collapse dramatically if no catch limits are set.

World catches of elasmobranchs are substantially higher than indicated by the different official statistics. Statistics reported to FAO amounted to just below 700 000t in 1991. Results presented here suggest that the total catch (as opposed to landings) is closer to 1 000 000t. This includes the estimated catch of the People's Republic of China and the by catch from large-scale high seas fisheries, but does not include discards from the bottom trawl fisheries around the world. Recreational fisheries are also not included since little information is available. However, they are important fisheries in many places, e.g. the USA, South Africa and Australia. Hoff and Musick (1990) estimate that the mortality of sharks in recreational fisheries of the eastern USA alone, is more than 10 000t/yr. The real total level of sharks, rays and chimaeras caught around the world is probably closer to 1 350 000t, twice official statistics!

3.2 Problems for the Assessment and Management of Elasmobranch Fisheries

3.2.1 Biology and Fisheries Theory

One of the chief problems in dealing with elasmobranch fisheries is that their biological and ecological characteristics make them particularly vulnerable to overexploitation. Most shark and many ray species can be classified as strong **K** strategists, i.e., they are long-lived with slow growth rates and late sexual maturation. Most species have low fecundity and these factors results in low reproductive potential. Further, they are usually the top predators in their communities (at least in the case of sharks) and thus have comparatively low abundances.

Some important areas of elasmobranch population dynamics are largely unknown. First, stock-recruitment relationships have never been demonstrated for any elasmobranch group though strong relationship is expected because of the reproductive strategies of the group. Second, there is a general lack of evidence about density-dependent mechanisms regulating elasmobranch population size. Third, the spatial structure and dynamics of most stocks are almost totally unknown. This is of particular importance to fisheries management at both the local and international level. Inadequate knowledge of migration routes, stock structure and movement-rates may undermine otherwise good assessments and management regimes. Much research, both practical and theoretical is still needed in these areas.

Another constraint to assessment and management of sharks and rays is inadequate population theory. For example, classical stock production models assume that there is an immediate response in the rate of population growth to changes in stock density, that the rate of natural increase at a given density is independent of the age composition of the stock and that exploited populations are in equilibrium. Neither of the first two assumptions seem to hold for elasmobranchs (Holden 1977, Wood et al. 1979); while the third probably does not hold for any fishery and surplus production models have been used for assessing of shark and ray fisheries without examination of the suitability of the model to the specific fishery. However, the difficulties in finding adequate models for elasmobranchs are exacerbated by the gaps in the understanding of their biology.

3.2.2 Multiplicity of Species and Gears

Additional problems for assessment and management are posed by the multispecific and multigear nature of most of the fisheries for sharks and rays. For example, elasmobranch catches in major tropical elasmobranch fishing countries account for 42% of reported world catches and are a mixture of several species of sharks, captured with a variety of gears from several types of vessels. Multispecies fisheries present difficult methodological problems because of the complexity of the biological and the technological interactions in the fisheries. Consequently, theoretical development of multispecies assessment and management still lags behind the rest of fisheries science (Hilborn and Walters 1992). In addition, the usage of multiple gears and fleets introduces difficulties in assessment and management, e.g. standardization of effort and allocation of quotas for the various types of gear and vessels.

3.2.3 Economics, Shark "finning" and Baseline Information

Many problems associated with elasmobranch exploitation are related to the economics of the fisheries. The economic processes involved in elasmobranch fisheries cause what could be called the "tragedy of sharks". This comes from two contradictory factors. First, research and management of sharks and rays are hampered by their low economic value: research funds are usually given to resources economically more important than elasmobranchs. Second, the high price attained by shark fins in the international market. This stimulates fisheries to target sharks and explains why incidental catches are usually "finned." The dynamics of the two processes means little hope for viable management consistent with both economic and conservation interests.

"Finning", i.e. cutting off the fins from the shark and dumping the carcass, is extremely wasteful but is common among fishermen throughout the world. Apart from being inhumane, finning is responsible for high death rates of sharks at sea. Finning is suspected to be particularly widespread in tuna fisheries but the extent and impacts of this habit are difficult to assess due to poor or non-existent information. This is another area where observer programmes of high seas fisheries could provide reliable information.

The low economic value of elasmobranchs results in fishery statistics which are not accurately maintained together with problems of species identification, specially for tropical species. Most records aggregate skates in a single group and sharks in two categories, large and small. Or even worse, the elasmobranchs are reported in a single category "various elasmobranchs". Without accurate statistics by species or species groups it is difficult to get insights into the dynamics of the stocks. Part of the answer to this problem lies in the economic field. When a specific market is developed for an elasmobranch species, catch statistics soon become available. Active development of markets for specific elasmobranch species may encourage better fishery statistics.

3.3 Conservation of Elasmobranchs

The top predator niche occupied by many sharks raises the question of their importance as regulators of other species' densities. Although it could be desirable to control shark populations in specific situations e.g., because they can affect the economy of important beach resort areas such a Natal or Hawaii, it is also possible that in other cases their removal would cause undesirable ecological and economical consequences (van der Elst 1979). It is difficult to assess these effects or to know which stocks of elasmobranchs are actually endangered when there is insufficient information about their ecology, size and state of their stocks, basic biology and the magnitude of their exploitation through fishing.

The size of the by catch of elasmobranchs in high seas fisheries is a major concern for conservation. Blue sharks might be facing extreme pressure in many parts of the globe but more specific studies are needed to determine the real situation. The threat that high seas fisheries pose to elasmobranchs is only one part of a complex interaction, e.g., There is substantial gear and catch damage caused by sharks in most of these fisheries (Taniuchi 1990, Sivasubramaniam 1963,1964, Pillai and Honma 1978, Berkeley and Campos 1988) and this causes financial loss for the fishing industries involved.

A solution to these problems could be to install shark deterrent devices in passive fishing gears (these account for most of the elasmobranch kill). The Natal Shark Board in South Africa is currently testing a electroacoustic device to protect bathers from shark attacks without having to kill the sharks. Another possibility is the design of selective fishing gear to reduce shark hooking rates. However, the only present viable alternative is the implementation of suitable by catch quotas for elasmobranchs in the high seas fisheries through international agreement and their enforcement through observer programmes.

The concern over elasmobranch exploitation arises from both theoretical considerations about their biological and ecological traits and for historical reasons. The record of fisheries for sharks and rays includes collapses and rapidly falling catch rates (Holden 1977). Examples include the California fishery for tope sharks, the piked dogfish fishery of British Columbia (1940's), the school shark fishery of Southern Australia (1950's), the porbeagle shark fishery in the Northwest Atlantic and the piked dogfish fishery in the North Sea (1960's) (Anderson 1990). Although the reasons for some of these collapses are partly understood and though decreasing CPUEs are a natural characteristic of fisheries development, these failures warn against high levels of exploitation in view of the special biological attributes of sharks and rays discussed above.

Protection of sharks and rays from the impacts of large-scale fisheries is not impossible. The efforts of international collaboration that regulated the catches of salmonids, marine birds and marine mammals in the North Pacific Ocean and the recent banning of all driftnet fisheries in the high seas of the world demonstrate it is possible. The strong pressure that some countries are imposing on fleets that take dolphins in purse seine tuna operations are another example that, where the will is there, protection becomes a reality.

Effective management and protection of elasmobranchs should begin with education and awareness. Only through intensive and widespread educational programmes is it possible to motivate fishermen, scientists, the public and governments to achieve effective protection and management of sharks and rays. Some of these efforts have already been successful. The South African Government has recently protected the great white shark; the Government of Australia forbids the killing of grey-nurse sharks and is considering protection of white sharks; California passed legislation banning the catch of great white sharks. During 1991, an international meeting, "Sharks Down Under", was held in Australia, focusing attention on the need for the conservation of elasmobranchs. The American Elasmobranch Society held a Symposium on Conservation of Elasmobranchs during its 1991 meeting and is presently establishing a Conservation Committee at the international level and the Species Survival Commission of the IUCN has recently formed a Shark Specialist Group. This is evidence of international concern about the future of elasmobranchs.

4. SUMMARY AND CONCLUSIONS

Elasmobranch fisheries are a traditional and common activity of minor global importance but they provide important sources of hard currency, protein and employment to many local communities. They are particularly important in Sri Lanka, Pakistan and Australia. Elasmobranch are fished with a range of gears from subsistence fisheries with artisanal gears and vessels, as is the case of some sail-powered boats in India, to highly industrialized fisheries with longlines, gillnets or trawls and as the distant water fishing nations of Japan, Taiwan (Prov. of China), Spain and the former Soviet Union.

There are 26 countries that are major exploiters of elasmobranchs (harvest more than 10 000t/yr). Among these, Japan, Indonesia, India, Taiwan (Prov. of China) and Pakistan have the highest average elasmobranch yields. About 30% of the 26 countries show recent falling trends in production. The analysis of Indexes of Relative Production by FAO major fishing areas suggests that further increases in exploitation of sharks and rays might possible, especially in the South East Pacific (Area 87), North East Pacific (Area 67) and the South East Atlantic (Area 47).

Although there are some specific fisheries for elasmobranchs (e.g. south Australian shark fishery, fisheries for sharks in Argentina and Mexico and basking shark fisheries of Norway), the larger part of world sharks and rays catches are taken incidentally. Official fisheries statistics do not properly reflect the amounts of sharks and rays harvested every year in the world's oceans. Although official figures report about 700 000t/yr of elasmobranchs caught at the end of the 1980's, the actual level is at least of 1 000 000t/yr and possibly 1 350 000t.

The by catches of sharks in large-scale high seas fisheries around the world are large, amounting possibly to almost 50% of the reported catches from commercial fisheries. The number of sharks caught annually in these fisheries during 1989-1991 is estimated at 11.6-12.7 million. The longline fisheries for tunas of Japan, Korea and Taiwan (Prov. of China) account for most of these bycatches. More detailed information is needed to address the magnitude of this problem and its effects upon shark populations. Observer programmes must be implemented for these fisheries to obtain reliable information about yields, discards, and the extent of finning practices. There are serious deficiencies in both the reporting and handling of the catch statistics. Of particular concern is the poor species discrimination which complicates appraisals. Fisheries statistics must be improved both in coverage of the fisheries and the dissaggregation of species.

5. BIBLIOGRAPHY

Alvarez, J.H. 1988. Analisis de la pesqueria de cazon de la peninsula de Yucatan. Centro de Investigacion y de Estudios Avanzados del Instituto Politecnico Nacional Unidad Merida, Mexico. 135 pp. (M.S. thesis)

Amorim, A.F. and C.A. Arfelli. 1987. Estudo biológico-pesqueiro do caçao-azul, *Prionace glauca*, no sudeste e sul do Brasil (1985-1986). III Reuniao do Grupo de Trabalho Sobre Pesca e Pesquisa de Tubaroes e Raias no Brasil, Fortaleza, Brasil. 28 July - 31 July, 1987. (Abstract)

Anderson, E.D. 1990. Estimates of large shark catches in the Western Atlantic and Gulf of Mexico, 1960-1986. NOAA Technical Report NMFS 90. pp. 443-454.

Anderson, E.D. and K. Teshima. 1990. Workshop on fisheries management. In: H.L. Pratt Jr., S.H. Gruber, and T. Taniuchi, (eds.) Elasmobranchs as living resources: advances in the biology, ecology, systematics, and the status of fisheries. NOAA Tech. Rep. NMFS 90.

Annala, J.H. (comp.). 1993. Report from the Fishery Assessment Plenary, May 1993: stock assessments and yield estimates. 241 p. Unpublished report held in MAF Fisheries Greta Point library, Wellington.)

Anon. 1983. New fishing ground located off Azhikode. Seafood Export J.; vol. 15, no. 6, p.8; 1983.

Anon. 1986. The types of fish caught, fishing grounds, and net classification. Yamaha Fishery Journal 1986. pp. 194

Anon. 1988. Report of the Secretary of Commerce to the Congress of the United States on the nature, extent, and effects of driftnet fishing in waters of the North Pacific Ocean, Pursuant to section 4005 of Public Law 100-220, the "driftnet impact monitoring, assessment and control act of 1987". 87p.

Anon. 1989. Status report on southern shark fishery. Australian Fisheries. 48(7), 14-19.

Anon. 1992a. Shark ban in the west. Australian Fisheries. 51(12):6

Anon. 1992b. Report for presentation to the united nations pursuant to resolutions 44/225 and 45/197. Scientific review of North Pacific high seas driftnet fisheries, Sidney B.C., June 11-14, 1991.

Appukittan K.K. and K.P. Nair. 1988. Shark resources of India, with notes on biology of a few species. Mangalore, Karnataka.; pp. 173-183.

Arfelli, C.A., A.F. Amorim and E.S. Rodrigues. 1987. Estudo biológico-pesqueiro do anequim, *Isurus oxyrinchus*, no sudeste e sul do Brasil (1971-1985). III Reuniao do Grupo de Trabalho Sobre Pesca e Pesquisa de Tubaroes e Raias no Brasil, Fortaleza, Brasil. 28 July - 31 July, 1987. (Abstract)

Aasen, O. 1963. Length and growth of the porbeagle (*Lamna nasus*, Bonaterre) in the North West Atlantic. FiskDir. Skr. Serie Havundersokelser 13(6):20-37.

Aasen, O. 1964. The exploitation of the spiny dogfish (*Squalus acanthias* L.) in European waters. FiskDir. Skr. Serie Havundersokelser 13:5-16.

Au, D.W.K. 1985. Species composition in the Japanese longline fishery off the southern and eastern United States. Coll. Vol. Sci. Pap. ICCAT 23(2):376-385.

Au, D.W.K. 1991. Polyspecific nature of tuna schools: shark, dolphin, and seabird associates. Fish. Bull. U.S. 89:343-354.

Bane, G.W. 1966. Observations on the silky shark, *Carcharhinus falciformis,* in the Gulf of Guinea. Copeia, 1966 (2):354-356

Batista, V.S. 1988. Determinação da Idade e Análise de Crescimento do Cação *Mustelus schmitti* Springer 1939 (Elasmobranchii, Triakidae). da Plataforma Continental do Rio Grande do Sul. M.Sc. Thesis. Univ. do Rio Grande, Brazil. 99 pp.

Bedford, D. 1987. Shark management: a case history -- the California pelagic shark and swordfish fishery. Sharks: an inquiry into biology, behaviour, fisheries and use. Proceedings of the Conference Portland, Oregon USA. October 13-15, 1985. pp. 161-172.

Berkeley, S.A. and W.L. Campos. 1988. Relative abundance and fishery potential of pelagic sharks along Florida's east coast. Marine Fisheries Review 50(1): 9-16.

Bernard, F.R. 1986. Data record: Flying squid drift-net and oceanographic cruise by W.E. Ricker, August-September 1986. (Document submitted to the Annual Meeting of the International North

Pacific Fisheries Commission, Anchorage, Alaska, November 1986) Department of Fisheries and Oceans, Fisheries Research Branch, Pacific Biological Station, Nanaimo, B.C., Canada V9R 5K6.

Bonfil, R. 1990. Contribution to the fisheries biology of the silky shark, Carcharhinus falciformis (Bibron 1839) from Yucatan, Mexico. MS Thesis. University College of North Wales, Bangor, UK. 77pp.

Bonfil, R. S., D. De Anda, and R. Mena. 1990. Shark Fisheries in Mexico: The case of Yucatan as an example. pp. 427-441. In: H.L. Pratt Jr., S.H. Gruber, and T. Taniuchi, (eds.) Elasmobranchs as living resources: advances in the biology, ecology, systematics, and the status of fisheries. NOAA Tech. Rep. NMFS 90.

Bonfil, R. (*in press*). Estado del conocimiento de los recursos de tiburones del Golfo de Mexico y Caribe. in: Analisis y diagnostico de los recursos pesqueros criticos del Golfo de Mexico. (eds. D. Flores-Hernandez, P. Sanchez-Gil, J.C. Seijo and F. Arreguín-Sanchez, Eds.) EPOMEX Serie Científica No. 4.

Brander, K.M. 1977. The management of Irish sea fisheries - a review. Laboratory leaflet No. 36. Lowestoff. 16 pp.

Brander, K.M. 1981. Disappearance of common skate *Raia batis* from Irish sea. Nature, vol 290. no. 5801: 48-49.

Cailliet, G.M., D.B. Holts, and D. Bedford, 1993. A review of the commercial fisheries for sharks on the West Coast of the United States. pp.13-29 in: Shark Conservation. Proceedings of an International Workshop on the Conservation of Elasmobranchs held at Taronga Zoo, Sydney, Australia, 24 February 1991. (Edited by Julian Pepperell, John West and Peter Woon). Zoological Parks Board of NSW.

Campbell, D., T. Battaglene, and S. Pascoe. 1991. Management options for the southern shark fishery. An economic analysis. Australian Bureau of Agricultural and Resource Economics. Australian Government Publishing Service, Canberra. 43 pp.

Castillo J.L.G. 1990. Shark fisheries and research in Mexico - a review. Chondros 2(1):1-2

Clark, M.R. and K.J. King. 1989. Deepwater fish resources off the North Island, New Zealand: results of a trawl survey, May 1985 to June 1986. New Zealand Fisheries Technical Report No. 11:56pp.

Coffey, B.T. and R.V. Grace. 1990. A preliminary assessment of the impacts of driftnet fishing on oceanic organisms: Tasman Sea, South Pacific, January 1990. Brian T. Coffey and Associates Limited: E.I.A./S.D.F. (G.P.01/1990). 19p. (MS)

Compagno, L.J.V. 1977. Phyletic relationships of living sharks and rays. Am. Zool., 17(2): 302-22.

Compagno, L.J.V. 1984. FAO species catalogue. Vol. 4. Sharks of the world. An annotated and illustrated catalogue of shark species known to date. Parts 1 and 2. FAO Fish. Synop. (125) Vol. 4:1-655.

Compagno, L. J. V. 1990. Shark exploitation and conservation. NOAA Tech. Rep. NMFS 90: 391-414.

Cook, S. F. 1990. Trends in shark fin markets: 1980, 1990 and beyond. Chondros. Vol. 2(1):3-6.

Crespo, E.A., and J.F. Corcuera. 1990. Interactions between marine mammals and fisheries in some areas of the coast of Argentina and Uruguay. IWC Document number SC/O90/G2: 31 pp.

Crummy, C., M. Ronan and E. Fahy. 1991. Blue shark *Prionace glauca* (L.) in Irish waters. Ir. Nat. J. vol. 23 (11):454-456.

Dahlgren, T. 1992. Shark longlining catches on, on India's east coast. Bay of Bengal's News, 48:10-12

Dayaratne, P. 1993a. Observations on the species composition of the large pelagic fisheries of Sri Lanka. Document presented at the 5th session of the Expert Consultation on Indian Ocean tunas, Mahé., Seychelles, 4-8 October, 1993. TWS/93/1/7. 8 pp.

Dayaratne, P. 1993b. Tuna fisheries in Sri Lanka - Present trends. Document presented at the 5th session of the Expert Consultation on Indian Ocean tunas, Mahé., Seychelles, 4-8 October, 1993. TWS/93/1/6. 10 pp.

Dayaratne, P., and R. Maldeniya. 1988. The status of tuna fisheries in Sri Lanka. Collective volume of documents presented at the Expert Consultation on stock assessment of Tunas in the Indian Ocean, Mauritius, 22-27 June, 1988. TWS/88/19: 292-303

Dayaratne, P., and J. de Silva. 1990. Drift gillnet fishery in Sri Lanka. Document presented at the Expert Consultation on stock assessment of Tunas in the Indian Ocean, Bangkok 2-6 July 1990. TWS/90/19. 7 pp.

De Metrio, G., G. Petrosino, C. Montanaro, A. Matarrese, M. Lenti, and E. Cecere. 1984. Survey on summer-autumn population of *Prionace glauca* L. (Pisces, Chondrichthyes) in the Gulf of Taranto (Italy) during the four year period 1978-1981 and its incidence on sword-fish (*Xiphias gladius*) and albacore (*Thunnus alalunga* (Bonn)) fishing. Oebalia, X:105-116

Departamento de Pesca. 1980a. Anuario estadistico pesquero 1978. Direccion general de planeacion, informatica y estadistica, Departamento de Pesca, Mexico. 361 pp.

Departamento de Pesca. 1980b. Anuario estadistico pesquero 1979. Direccion general de planeacion, informatica y estadistica, Departamento de Pesca, Mexico. 442 pp.

Departamento de Pesca. 1981. Anuario estadistico pesquero 1980. Direccion general de planeacion, informatica y estadistica, Departamento de Pesca, Mexico. 800 pp.

Departamento de Pesca. 1982. Anuario estadistico pesquero 1981. Direccion general de planeacion, informatica y estadistica, Departamento de Pesca, Mexico. 796 pp.

Devadoss, P. 1978. On the food of rays, *Dasyatis uarnak, D. alcockii* and *D. sephen.* Indian J. Fish., 25:9-13.

Devadoss, P. 1984. On the incidental fishery of skates and rays off Calicut. Indian J. Fish., vol. 31: 285-292.

Devadoss, P. 1988. Observations on the breeding and development of some sharks. J. Mar. Bio. Assoc. India., vol. 30, no.1-2, pp. 121-131.

Devadoss, P., M.D.K. Kuthalingam and R. Thiagarajan. 1988. Present status and future prospects of elasmobranch fishery in India. CMFRI Spec. Publ.; no. 40, pp. 29-30.

Devaraj, M., and P. Smita. 1988. Economic performance of mechanised trawlers in the state of Kerala, India. Fish. Res., vol. 6, no. 3, pp. 271-286.

Encina, V.B. 1977. The discovery and distribution of the spiny dogfish shark resource in The Philippines. The Philippine Journal of Fisheries, 11(1-2):127-139

Evangelista, J.E.V. 1987. Desembarque artesanal de caçoes nos municípios de Belém (Ver-O-Peso), Marapanum (Mercado municipal), Primavera (Sao Joao de Pirabas) e Vigia (Mercado municipal e apapiranga), durante os anos de 1983 a 1986 e estimativa da captura de caçoes e raias na pesca industrial de piramutaba (*Brachyplatystoma vaillanti*) no esteário do rio Amazonas, durante os anos de 1982 a 1986. III Reuniao do Grupo de Trabalho Sobre Pesca e Pesquisa de Tubaroes e Raias no Brasil, Fortaleza, Brasil. 28 July - 31 July, 1987. (Abstract)

Fahy, E. 1989a. The spurdog *Squalus acanthias* (L) fishery in south west Ireland. Irish Fisheries Investigations. Series B (Marine) No. 32, 1988. 22 pp.

Fahy, E. 1989b. Fisheries for Ray (Batoidei) in Western statistical area viia, investigated through the commercial catches. Irish Fisheries Investigations. Series B (Marine) No. 34, 1989. 14 pp.

Fahy, E. 1991. The south easter ray *Raja* spp. fishery, with observations on the growth of rays in Irish waters and their commercial grading. Irish Fisheries Investigations. Series B (Marine) No. 37, 1991. 18 pp.

Fahy, E. and P. Gleeson. 1990. The post-peak-yield gillnet fishery for spurdog *Squalus acanthias* L. in Western Ireland. Irish Fisheries Investigations. Series B (Marine) No. 35. 12 pp.

FAJ (Fisheries Agency of Japan). 1987-1989. Data record of fishes and squids caught incidentally in gillnets of Japanese salmon research vessels, 1986-1988. (Document submitted to the International North Pacific Fishery Commission). (MS).

FAJ (Fisheries Agency of Japan). 1983. Survey on the feasibility of using surface gillnets for pomfret and driftnets for flying squid resources in the North Pacific Ocean in 1982. (Document submitted to the International North Pacific Fishery Commission)

FAJ (Fisheries Agency of Japan). 1989. Summary of observations for Japanese squid driftnet fishery in the north Pacific in 1988. (Document submitted to the International North Pacific Fishery Commission). (MS).

FAJ (Fisheries Agency of Japan). 1990. Salmon catch statistics for Japanese mothership gillnet and landbased fisheries 1989. (Document submitted to the International North Pacific Fishery Commission). (MS).

FAJ (Fisheries Agency of Japan). 1991. Salmon catch statistics for Japanese non-traditional and traditional landbased salmon fisheries, 1991. 24p. Fisheries Agency of Japan. Kasumi-gaseki 1-2-1, Chiyoda-ku, Tokyo, Japan 100.

FAO (Food and Agriculture Organization of the United Nations). 1990. Yearbook of Fishery Statistics: catches and landings 1988, Vol 66. Food and Agriculture Organization of the United Nations. Rome.

FAO (Food and Agriculture Organization of the United Nations). 1991. Yearbook of Fishery Statistics: catches and landings 1989, Vol. 68. Food and Agriculture Organization of the United Nations. Rome.

FAO (Food and Agriculture Organization of the United Nations). 1992. Yearbook of Fishery Statistics: catches and landings 1990, Vol 70. Food and Agriculture Organization of the United Nations. Rome.

FAO (Food and Agriculture Organization of the United Nations). 1993. Yearbook of Fishery Statistics: catches and landings 1991, Vol 72. Food and Agriculture Organization of the United Nations. Rome.

Fitzgerald, S.M., H. McElderry, H. Hatanaka, Y. Watanabe, J.S. Park, Y. Gong and S.Y. Yeh. 1993. 1990-1991 North Pacific high seas driftnet scientific observer programs. International North Pacific Fisheries Commission. Bulletin no. 53(I): 77-90.

Francis, M.P. 1989. Exploitation rates of rig (*Mustelus lenticulatus*) around the South Island of New Zealand. N. Z. J. Mar. Freshwater Res. 23:239-245

Francis, M.P. and R.I.C.C. Francis. 1992. Growth rate estimates for New Zealand rig (*Mustelus lenticulatus*). 1157-76 **in:** "Age determination and growth in fish and other aquatic animals". (ed. D.C. Smith). Aust. J. of Mar. and Freshwater Res 43(5).

Francis, M.P. and J.T. Mace. 1980. Reproductive biology of *Mustelus lenticulatus* from Kaikoura and Nelson. N. Z. J. Mar. Freshwater Res. 14(3):303-311

Francis, M.P. and D.W. Smith. 1988. The New Zealand rig fishery: catch statistics and composition, 1974-85. New Zealand Fisheries Technical Report No. 7. 30 pp.

Galván-Magaña, F., H.J. Nienhuis and A.P. Klimley. 1989. Seasonal abundance and feeding habits of sharks of the lower Gulf of California, Mexico. Cal. Fish and Game 75(2):74-84

Garces, A., and J.C. Rey. 1984. La pesquería espanola de pez espada (*Xiphias gladius*), 1973-1982. Col. Vol. Sci. Pap. Vol. XX, (2): 419-427.

Gauld, J.A. 1989. Records of porbeagles landed in Scotland, with observations on the biology, distribution and exploitation of the species. Scottish Fisheries Research Report 45: 15 pp.

Gjernes, T., S, McKinnell, A. Yatsu, S. Hayase, J. Ito, K, Nagao, H. Hatanaka, H. Ogi, M. Dahlberg, L. Jones, J. Wetherall and P. Gould, 1990. Final report of squid and by-catch observations in the Japanese driftnet Fishery for neon flying squid (*Ommastrephes bartrami*) June-December, 1989. Observer Program. Joint report of Fisheries Agency of Japan, Canadian Departments of Fisheries and Oceans, United States National Marine Fisheries Service, United States Fish and Wildlife Service. 114 p.

GMFMC (Gulf of Mexico Fisheries Management Council). 1980. Fishery management plan for the shark and other elasmobranch fishery of the Gulf of Mexico. GMFMC. Tampa, Florida.

Göcks, W.E.G. 1987. Industrialização Total Dos Tubaroes e Caçoes. III Reuniao do Grupo de Trabalho Sobre Pesca e Pesquisa de Tubaroes e Raias no Brasil, Fortaleza, Brasil. 28 July - 31 July, 1987. (Abstract)

Gong, Y., J.U. Lee and Y.S. Kim. 1987. Fishing efficiency of Korean deep longline gear and vertical distribution of tunas in the Indian Ocean. Collective volume of working documents presented at the expert consultation on stock assessment of tunas in the Indian Ocean held in Colombo, Sri Lanka, 4-8 December 1986.

Gong, Y., J.U. Lee, Y.S. Kim and W.S. Yang. 1989. Fishing efficiency of Korean regular and deep longline gears and vertical distribution of tunas in the Indian Ocean. Bull. Korean. Fish. Soc. 1989. vol.22 (2): 86-94.

Gong, Y., S.K. Yeong, and S.J. Hwang. 1993. Outline of the Korean squid gillnet fishery in the North Pacific. I.N.P.F.C. Bulletin 53(1): 45-70

Grulich, P.R. and W.D. DuPaul. 1987. Development considerations for the dogfish (*Squalus acanthias*) fishery of the Mid-Atlantic. pp. 201-214, **in:** Sharks: an inquiry into biology, behaviour, fisheries and use. Proceedings of the Conference Portland, Oregon USA. October 13-15, 1985. (edited by S. Cook)

Hanchet, S. 1988. Reproductive biology of *Squalus acanthias* from the east coast, South Island, New Zealand. N. Z. J. Mar. Freshwater Res. 22:537-549

Hanchet, S. 1991. Diet of spiny dogfish, *Squalus acanthias* Linneaus, on the east coast, South Island, New Zealand. N. Z. J. Mar. Freshwater Res. 39(3):313-323

Hazin, F.H., A.A. Couto, K. Kihara, K. Otsuka, and M. Ishino. 1990. Distribution and abundance of pelagic sharks in the south-western equatorial Atlantic. J. Tokyo Univ. Fish. 77(1):51-64

Hilborn, R. and C. J. Walters. 1992. Quantitative Fisheries Stock Assessment: Choice, dynamics and uncertainty. Chapman and Hall. 570 pp.

Ho, P.S.D. 1988. Development of Taiwanee Tuna Fishery, pp 209-210 in "Report of the 2nd Southeast Asian Tuna Conference and 3rd meeting of tuna resarch groups in the southeast Asia Region". Kuala Terengganu, Malyasia, 22-25 August 1988. IPTP/88/GEN15. 220 pp.

Hoff, T.B. 1990. Conservation and management of the western north Atlantic shark resource based on the life history strategy limitations of sandbar sharks. Ph. D. Dissertation, University of Delaware, 282 pp.

Hoff, T.B. and J.A. Musick. 1990. Western North Atlantic shark-fishery management problems and informational requirements. pp. 455-472.

Holden, M.J. 1968. The rational exploitation of the Scottish-Norwegian stocks of spurdogs (*Squalus acanthias* L.). Fisheries Investigations, series II, vol.24(8). 27 pp.

Holden, M.J. 1977. Elasmobranchs. pp 187- 215. In: J. A. Gulland (Ed), Fish Population Dynamics . J. Wiley and Sons, Lond.

Holden M.J. and P.S. Meadows. 1962. The structure of the spine of the spurdogfish (*Squalus acanthias* L.) and its use for age determination. J.Mar.Biol.Ass. U.K., 42:179-97.

Holden M.J. and P.S. Meadows. 1964. The fecundity of the spurdog (*Squalus acanthias* L.). J.Cons.perm.int.Explor.Mer, 28:418-24.

Holts, D.B. 1988. Review of U.S. West Coast Commercial Shark Fisheries. Marine Fisheries Review. 50(1). pp. 1-8.

Hsu, C. and H. Liu. 1990. Taiwanese longline and gillnet fisheries in the Indian (sic). Paper submitted to the Expert Consultation on Stock Assessment of Tunas in The Indian Ocean. Bangkok, Thailand, 2-6 July 1990. (MS).

Hsu, C. and H. Liu. 1991. Taiwanese longline and gillnet fisheries in the Indian Ocean. Coll. Vol of Working Documents Vol. 4, Presented at the Expert Consultation on Stock Assessment of Tunas in The Indian Ocean Held in Bangkok, Thailand, 2-6 July 1990:244-260

Hsu, C. and H. Liu. 1992. Status of Taiwanese longline fisheries in the Atlantic (sic). ICCAT Coll. Vol. Sci. Paps. 39 (1):258-264

Hsu, C. and H. Liu. 1993. Status of Taiwanese longline fisheries in the Atlantic (sic), 1991. ICCAT Coll. Vol. Sci. Paps. 40(2):330-332

ICCAT (International Commission for the Conservation of Atlantic Tunas). 1991a. Data Record. Vol. 32:387 pp.

ICCAT (International Commission for the Conservation of Atlantic Tunas). 1991b. Report for biennial period 1990-1991. Part I (1990): 455 pp.

ICCAT (International Commission for the Conservation of Atlantic Tunas). 1992. Data Record. Vol. 33:285 pp.

ICCAT (International Commission for the Conservation of Atlantic Tunas). 1993a. Report of the second GFCM/ICCAT expert consultation on stocks of large pelagic fishes in the Mediterranean Sea. pp 11-35 **In**: Collective Volume of Scientific Papers 40(1). ICCAT, Madrid, Spain.

ICCAT (International Commission for the Conservation of Atlantic Tunas). 1993b. The Report of the Data Preparatory Meeting for South West Atlantic Tuna and Tuna-like Fisheries. Recife, Pernambuco, Brazil, 1-7 July 1992. pp. 1-19 **In**: Collective Volume of Scientific Papers 40(2). ICCAT, Madrid, Spain.

INPFC (International North Pacific Fisheries Commission). 1990. Final report of squid and by-catch observations in the Japanese driftnet fishery for neon flying squid (*Ommastrephes bartrami*). June-Dec., 1989. Observer programme joint report by the National Sections of C.J.U.

INPFC (International North Pacific Fisheries Commission). 1991. Final report on 1990 observations on the Japanese high seas squid driftnet fisheries in the North Pacific Ocean. Joint report by the National sections of Canada, Japan and the United States. 198 pp.

INPFC (International North Pacific Fisheries Commission). 1992. Final report of observations of the Japanese high seas large-mesh driftnet fishery in the North Pacific Ocean, 1990-1991, Joint report by the National Sections of Canada, Japan and the United States. 86 pp.

INPFC (International North Pacific Fisheries Commission). 1993. Statistical Yearbook 1990. 116 pp.

IPTP (Indo-Pacific Tuna Development and Management Programme). 1989. Tuna sampling programme in Sri Lanka. IPTP/89/SRL/SP. 109 pp.

IPTP (Indo-Pacific Tuna Development and Management Programme). 1990. Report of the Expert Consultation on Stock Assessment of Tunas in The Indian Ocean. Bangkok, Thailand, 2-6 July 1990. 96 pp.

IPTP (Indo-Pacific Tuna Development and Management Programme). 1991. Tuna sampling programme in Karachi, Pakistan. IPTP/91/PAK/SP. 45 pp.

Ishihara H. 1990. The skates and rays of the western North Pacific: an overview of their fisheries, utilization, and classification. NOAA Tech. Rep. NMFS 90: 485-498.

Ito, J., S. Hayase, and A. Yatsu. 1990. Outline of the experimental survey of the subsurface squid gillnets by Kanki-Maru No. 3 in the North Pacific in Sept/Oct 1989. National Research Institute of Far Seas Fisheries.

Ito, J., W. Shaw, and R.L. Burgner. 1993. Symposium on biology, distribution and stock assessment of species caught in the high seas driftnet fisheries in the North Pacific Ocean. International North Pacific Fisheries Commission, Bulletin no. 53(I,II,III).

Ivanov,L. and R.J.H. Beverton. 1985. The fisheries resources of the Mediterranean. Part Two: Black Sea. Etud, Rev. CGPM/Stud, Rev. GFCM, (60);135pp.

Jacinto, M.A.C. 1987. Aspectos do curtimento de pele de tubarao. III Reuniao do Grupo de Trabalho Sobre Pesca e Pesquisa de Tubaroes e Raias no Brasil, Fortaleza, Brasil. 28 July - 31 July, 1987. (Abstract)

James, P.S.B.R. 1988. Recent observations on marine fisheries resources of Lakshadweep. Mangalore, Karnataka.; 191-197 pp.

James, P.S.B.R., and A.A. Jayaprakash. 1988. Status of tuna fishing in India. pp. 278-291 in: Collective Volume of Working Documents Presented at the Expert Consultation on stock Assessment of tunas in the Indian Ocean held in Mauritius, 22-27 June 1988.

James, P.S.B.R., and P.P. Pillai. 1987. A review of national tuna fishery - India. pp. 342-352 in: Collective Volume of Working Documents Presented at the Expert Consultation on stock Assessment of tunas in the Indian Ocean held in Colombo, Sri Lanka 4-8 December 1986.

King, K. and M. Clark. 1987. Sharks from the upper continental slope -- are they of value? Catch. May 1987. pp. 3-6.

Kirkwood, G.P. and T.I. Walker. 1986. Gill net mesh selectivities for Gummy shark, *Mustelus antarcticus* Günther, taken in south-eastern Australian waters. Aust. J. Mar. Freshw. Res., 1986, 37: 689-97.

Krishnamoorthi, B. and I. Jagadis. 1986. Biology and population dynamics of the grey dogshark *Rhizoprionodon (Rhizoprionodon) acutus* (Ruppel) in Madras waters. Indian J. Fish; 33(4): 371-385.

Kulkorni, G.N. and S.T. Sharangdher. 1990. Disco net fishing off Ratnagiri coast. Fish Games, vol. 9, no. 12, pp. 37-39, 41.

Kunjipalu K.K. and A.C. Kuttappan 1978. Note on an abnormal catch of devil rays *Dicerobatis eregoodoo* Day in gill nets off Veraval. Indian J. Fish.; vol. 25, no. 1-2, pp. 254-256.

Kunslik, P.A. 1988. The basking shark. Scottish Fisheries Information Pamphlet No. 14. Department of Agriculture and Fisheries of Scotland. 21 pp.

Lablache, G. and B. Karpinski. 1988. Seychelles research observer program. IPTP Collective Volume of Working Documents Vol. 3. Presented at the expert consultation on stock assessment of tunas in the Indian Ocean held in Mauritius, 22-27 June 1988:146-153

Lawlor, F. and S. Cook. 1987. U.S. shark fishing methods and gear. Sharks: an inquiry into biology, behaviour, fisheries and use. Proceedings of the Conference Portland, Oregon USA. October 13-15, 1985. pp. 137-146.

Lawson, T. 1991. Status of tuna fisheries in the SPC area during 1990, with annual catches since 1952. South Pacific Commission, Tuna and Billfish Assessment Programme Tech. Rep. 27:73pp.

LeBrasseur, R., B. Riddell, and T. Gjernes. 1987. Ocean salmon studies in the Pacific subarctic boundary area. 16 p. (Document submitted to the annual meeting of the International North Pacific Fishery Commission, Vancouver, Canada, October 1987). Department of Fisheries and Oceans, Biological Sciences Branch, Pacific Biological Station, Nanaimo, B.C. V9R 5K6.

Lenanton, R.C.J., D. I. Heald, M. Platell, M. Cliff, and J. Shaw. 1990. Aspects of the reproductive biology of the gummy shark, *Mustelus antarcticus* Günter, from waters off the south coast of Western Australia. Aust. J. of Mar. and Freshwater Res. 41:807-22

Lopez, A.M., D.B. McClellan, A.R. Bertolino and M.D. Lange. 1979. The Japanese longline fishery in the Gulf of Mexico, 1978. Mar. Fish. Rev. 41(10):23-28

Lyle, J.M. 1984. Mercury concentrations in four carcharhinid and three hammerhead sharks from coastal waters of the Northern Territory. Aust. J. Mar. Freshw. Res., 1984. 35: 441-51.

Martin L. and G.D. Zorzi, 1993. Status and review of the California skate fishery. NOAA Technical Report NMFS 115:39-52

Massey, B.R., and M.P. Francis. 1989. Commercial catch composition and reproductive biology of rig (*Mustleus lenticulatus*) from Pegasus Bay, Canterbury, New Zealand. N. Z. J. Mar. Freshwater Res. 23:113-120

Maxwell, G. 1952. Harpoon at a venture. London, Rupert Hart-Davis, 272 pp.

McKinnell, S.M., T. Gjernes, W. Shaw and S. Whiteaker. 1989. Canadian North Pacific Pelagic Study, Arctic Harvester, July 12 - August 22, 1989. 19 p. (Document submitted to the Annual Meeting of the International North Pelagic Fisheries Commission, Seattle, Washington, U.S.A., October 1989). Department of Fisheries and Oceans, Pacific Biological Station, Nanaimo, B.C., Canada V9R 5K6.

Mejuto, J. 1985. Associated catches of sharks, *Prionace glauca, Isurus oxyrinchus,* and *Lamna nasus,* with NW and N Spanish swordfish fishery, in 1984. ICES C.M.1985/H:42: 16 pp.

Mejuto, J., B. García and J.M. de la Serna. 1993. Activity of the spanish surface longline fleet targeting swordfish (*Xiphias gladius* L.) in the Atlantic, years 1988 to 1991, combined. ICCAT Coll. Vol. Sci. Paps. 40(1):393-399

Mejuto, J. and S. Iglesias. 1988. Campaña comercial de prospección de abundancia de pez espada, *Xiphias gladius* L., y especies asociadas, en areas próximas a Grand Banks. ICCAT Coll. Vol. Sci. Paps. 27:155-163

Menasveta, D., S. Shino and S. Chullasorn. 1973. Pelagic fishery resources of the South China Sea and prospects for their development. UNDP/FAO. SCS/DEV/73/6. 68pp.

Menni, R.C., MB.Cousseau, and A.R. Gosztonyi. 1986. Sobre la biologia de los tiburones costeros de la provincia de Buenos Aires. Anales de la Sociedad Cientifica Argentina 213:3-27

Millington, P.J. 1981. The Taiwanese gillnet fishery in the Australian Fishing Zone: A preliminary analysis of the first year's operation. pp. 119-144 **in:** Grant, C.J. and D.G. Walter (eds.) "Northern Pelagic Fish Seminar (Darwin 1981)". Department of Primary Industry, Australian Government Publishing Service. Canberra.

Mimura, K et al. 1963. Synopsis of biological data on yellowfin tuna, *Neothunnus macropterus* Temminch and Schlegel, 1842. Species synopsis No. 10, FAO Fish. Rep 6(2):319-349

Moulton, P.L., T.I. Walker, and S.R. Saddlier. 1992. Age and growth of gummy shark, *Mustelus antarcticus* Günter, and school shark, *Galeorhinus galeus* (Linneaus), from southern Australian waters. pp. 879-911 **in:** "Age determination and growth in fish and other aquatic animals". (ed. D.C. Smith). Aust. J. of Mar. and Freshwater Res 43(5):

Muñoz-Chápuli, R. 1985a. Analisis de las capturas de escualos demersales en el Atlantico NE (27°N-37°N) y mar de Alboran (Mediterraneo Occidental). Investigacion Pesquera 49(1):121-136

Muñoz-Chápuli, R. 1985b. Analisis de las capturas de escualos pelagicos en el Atlantico nororiental (15°-40°N). Investigacion Pesquera 49(1):67-79

Murata, M. 1986. Report on fishing survey on flying squid by the drift gillnetters Shoyo Maru, Kuromori Maru No 38 and Kanki Maru No. 58 in the North Pacific in 1985. Fisheries Agency of Japan. INPFC.

Murata, M. 1987. Report on fishing survey on flying squid by the drift gillnetters Shoyo Maru, Kuromori Maru No 38 and Kanki Maru No. 58 in the North Pacific in 1986. (Document submitted to the Annual Meeting of the International North Pacific Fisheries Commission, Vancouver, Canada, 1987, October). 20 p. Fisheries Agency of Japan, Tokyo, Japan 100.

Murata, M., Y. Nakamura and H. Saito. 1989. Report on fishing survey on flying squid by Shoyo Maru, Kanki Maru No. 3 and Hoyo-Maru No. 78 in the North Pacific in 1988. (Document submitted to the Annual Meeting of the International North Pacific Fisheries Commission, Seattle, Washington, 1989 October). 25 p. Fisheries Agency of Japan, Hokkaido Regional Fisheries Research Laboratory, 116 Katsurakoi, Kushiro, Japan 085.

Murata, M. and C. Shingu. 1985. Report on fishing survey on flying squid by drift gillnetters Oumi-Maru and Kanki-Maru No. 58 in the North Pacific in 1984. Fisheries Agency of Japan. INPFC.

Murray, T. 1990. Review of research and of recent developments in South Pacific albacore fisheries, with emphasis on large-scale pelagic driftnet fishing. Third South Pacific Albacore Research Workshop. Inf. Pap. No.2. SPC. Noumea, New Caledonia 9-12 Oct. 1990. 27 pp.

Myers, K.W., C.K. Harris, Y. Ishida, L. Margolis and M. Ogura. 1993. Review of the Japanese landbased driftnet salmon fishery in the Western North Pacific Ocean and the continent of origin of salmonids in this area. International North Pacific Fisheries Commission. Bulletin no. 52. 86 pp.

Myklevoll, S. 1989a. Norway's porbeagle fishery. Working document presented at the ICES Study Group on Elasmobranch Fisheries, Dublin Ireland, 26-28 April 1989. (mimeo).

Myklevoll, S. 1989b. Norway's basking shark fishery. Working document presented at the ICES Study Group on Elasmobranch Fisheries, Dublin, Ireland, 26-28 April 1989. (mimeo).

Myklevoll, S. 1989c. Norway's greenland shark fishery. Working document presented at the ICES Study Group on Elasmobranch Fisheries, Dublin, Ireland, 26-28 April 1989. (mimeo).

Myklevoll, S. 1989d. Norway's spurdog fisheries. Working document presented at the ICES Study Group on Elasmobranch Fisheries, Dublin, Ireland, 26-28 April 1989. (mimeo).

Myklevoll, S. 1989e. Norway's skate and ray fishery. Working document presented at the ICES Study Group on Elasmobranch Fisheries, Dublin, Ireland, 26-28 April 1989. (mimeo).

Nagao, K., S. Ota, and J. Hirono. 1993. Regulation of the Japanese high seas driftnet fisheries. International North Pacific Fisheries Commission. Bulletin no. 53(I): 39-44.

Nakano, H. 1993. A review of the Japanese fishery and research on sharks in the Atlantic Ocean. ICCAT Coll. Vol. Sci. Pap. 40(2):409-412

Nakano, H., M. Makihara, and K. Shimazaki. 1985. Distribution and biological characteristics of the blue shark in the central north Pacific. Bull. Faculty Fish., Hokkaido Univ. 36(3) 99-113

Nakano, H., K, Okada, Y, Watanabe, and K. Uosaki. 1993. Outline of the large-mesh driftnet fishery of Japan. International North Pacific Fisheries Commission Bulletin No. 53 (I): 25-38

Nakano, H. and Y. Watanabe. 1992. Effect of high seas driftnet fisheries on blue shark stock in the North Pacific. 15 pp. Compendium of documents submitted to the Scientific Review of North Pacific Highseas Driftnet Fisheries, Sidney, B.C., Canada, June 11-14, 1991, Vol. 1.

NFRDA (National Fisheries Research and Development Agency of Korea). 1988. Annual report of catch and effort statistics and fishing grounds for the Korean tuna longline fishery 1983-1985. 610 pp.

NFRDA (National Fisheries Research and Development Agency of Korea). 1992. National Report of Korea. pp. 427-429 in: ICCAT Report for biennial period, 1990-1991 Part I (1990): 455 pp.

NMFS (National Marine Fisheries Service). 1993. New rules regulate shark fishing in US Atlantic, Gulf and Caribbean waters. News Bulletin NR93-15. NMFS Southeast Regional Office, St. Petersburg, FL.

NOAA (National Oceanic and Atmospheric Organization). 1991. Fishery management plan for sharks of the Atlantic Ocean. NMFS, NOAA. U.S. Department of Commerce. October 28, 1991. 147 pp.

Northridge, S.P. 1991. Driftnet fisheries and their impacts on non-target species: a worldwide review. FAO Fisheries Technical Paper, 320:115p.

NRIFSF (National Research Institute of Far Seas Fisheries of Japan). 1992. National Report of Japan. pp. 423-426 in: ICCAT Report for biennial period, 1990-1991 Part I (1990): 455 pp.

O'Connor, P.F. 1953. Shark-O! London, Secker and Warburg, 272 pp.

Okera, W., J.D. Stevens and J.S. Gunn. 1981. Fishery situation report: tropical sharks. Northern Pelagic Fish Seminar (Darwin 1981). Department of Primary Industry. Fisheries Division Australian Government Publishing Service. Canberra. 177 pp.

Pajot, G. 1980. Improvement of large-mesh driftnets for small scale fisheries in Bangladesh. FAO Bay of Bengal Programme Doc. BOBP/WP/5. 14p.

Parrack, M.J. 1990. A study of shark exploitation in U.S. Atlantic coastal waters during 1986-1989. Contribution MIA-90:91-03. NOAA, NMFS, Miami, Florida. 14 pp.

Paust, B.C. 1987. The developing Alaska salmon shark fishery. in: "Sharks an inquiry into biology, behaviour, fisheries, and use" (S. Cook ed.), pp. 173-178. Oregon State University Extension Service.

Pella, J. R. Rumbaugh, L. Simon, M. Dahlberg, S. Pennoyer, and M. Rose. 1993. Incidental and illegal catches of salmonids in the North Pacific driftnet fisheries. I.N.P.F.C. Bulletin 53(3):325-358

Peres, M. B. and C. M. Vooren. 1991. Sexual development, reproductive cycle, and fecundity of the school shark *Galeorhinus galeus* off southern Brazil. Fishery Bull. U.S. 89:655-667

Pillai, P.P. and M. Honma. 1978. Seasonal and areal distribution of the pelagic sharks taken by the tuna longline fishery in the Indian Ocean. Bulletin of the Far Seas Fisheries Research Laboratory No. 16:33-48.

Pope, J. 1979. Stock assessment in multispecies fisheries, with special reference to the trawl fishery in the Gulf of Thailand. UNDP/FAO. SCS/DEV/79/19, 121 pp.

Prado, J., and S. Drew. 1991. Trials and developments in small scale shark fishing carried out by FAO, 1978-1990. FAO Fisheries Circular No. 840. 68 pp.

Pratt, H.L. 1979. Reproduction in the blue shark, *Prionace glauca*. Fish. Bull. 77:445-70

Pratt, H.L., and J.G. Casey. 1990. Shark reproductive strategies as a limiting factor in directed fisheries, with a review of Holden's method of estimating growth parameters. In: H.L. Pratt Jr., S.H. Gruber, and T. Taniuchi, (eds.) Elasmobranchs as living resources: advances in the biology, ecology, systematics, and the status of fisheries. NOAA Tech. Rep. NMFS 90: 97-110

Rama R., S.V.S., P.G. Mathai, V.C. George, K.K. Kunjipalu, M.D. Varghese and A.C. Kuttappan. 1989. Shark long line gear of India. Fish. Technol. Soc. Fish. Technol., Cochin.; vol. 26, no. 2, pp. 73-80.

Reuben, S., G.S. Rao, G. Luther, T.A. Rao, K. Radhakrishna, Y.A. Sastry, and G. Radhakrishnan. 1988. An assessment of the demersal fishery resources of the northeast coast of India. CMFRI spec. publ.; no.40, p. 15.

Rey, J.C., and E. Alot. 1984. Contribucion al estudio de la pesquería de palangre de pez espada (*Xiphias gladius*) en el Mediterraneo Occidental. Col. Vol. Sci. Pap. Vol. XX, (2): 428-434.

Rey, J.C., and R. Muñoz-Chápuli. 1991. Relation between hook depth and fishing efficiency in surface longline gear. Fish. Bull. U.S. 89:729-732

Richard, J. 1987. Developing a localized fishery: the Pacific angelshark. Sharks: an inquiry into biology, behavior, fisheries and use. Proceedings of the Conference Portland, Oregon USA. October 13-15, 1985. pp. 147-160.

Ripley, W.E. 1946. "The soupfin shark and the fishery" in the Biology of the Soupfin *Galeorhinus zyopterus* and Biochemical Studies of the Liver. Fish Bulletin No. 64. California Division of Fish and Game. pp. 7-37.

Robertson, D.A., P.J. Grimes, and P.J. McMillan. 1984. Orange roughy on Chatham Rise; results of a trawl survey, August-September 1982. Fisheries Research Division, Occasional Publication No. 46: 27pp.

Rodriguez, A., S.F. Nieto, and L. Muñoz. 1988. Análisis de la abundancia (1973-1985) de grandes peces pelágicos en la zona oceánica del Atlántico tropical-oriental. ICCAT Coll. Vol. Sci. Paps. 38:339-349

Roedel, P.M. and W.E. Ripley. 1950. California sharks and rays. Fish Bulletin No. 75. California Division of Fish and Game. 29 pp.

Rohan, G.V. 1981. The market situation for tropical pelagic fish in Australia. Northern Pelagic Fish Seminar (Darwin 1981). Department of Primary Industry. Fisheries Division Australian Government Publishing Service. Canberra. 177 pp.

Ross, A, and K. Bailey. 1986. The foreign longline catch of marlins, swordfish, and mako shark in the New Zealand EEZ for 1984. Catch 13(4-5):12-16

Rowlett, R.A. 1988. Cruise report. Republic of Korea research vessel Dusan 851. Survey of flying squid resources in the Northwestern Pacific Ocean. July-Aug. 1988. U.S. Dept. com. NMFS. Northwest and Alaska Fish. Center. Seattle, Washington.

Russell. S.J. 1991. Observations on the directed and by-catch longline fisheries for sharks in the Northern Gulf of Mexico. American Society of Ichthyologists and Herpetologists, 7th Annual Meeting, 15-20 June 1991. American Elasmobranch Society. (Abstract)

Ryland, J.S. and T.O. Ajayi. 1984. Growth and population dynamics of three *Raja* species (Batoidei) in Carmarthen Bay, British Isles. J. Cons. int. Explor. Mer, 41: 111-120.

Saika, S. and Yoshimura, H. 1985. Oceanic whitetip shark (*Carcharhinus longimanus*) in the western Pacific. Rep. Jap. Group for Elasmobranch Studies 20:11-21

Sakagawa, G.T., and P.M. Kleiber. 1992. Fisheries and stocks of yellowfin tuna in the Pacific and Indian Oceans: status and review of assessment methods. Collect. Vol. Sci. Pap. Iccat. vol. 38, pp. 203-217.

Santhanakrishnan, G. 1983. A diversified product and its export potential from India. Seafood Export J.; vol. 15, no. 2, pp. 19-24.

Secretaria de Pesca. 1984. Anuario estadistico de Pesca 1982. Direccion general de informatica y estadistica, Secretaria de Pesca, Mexico. 513 pp.

Secretaria de Pesca. 1985a. Anuario estadistico de Pesca 1983. Direccion general de informatica y estadistica, Secretaria de Pesca, Mexico. 327 pp.

Secretaria de Pesca. 1985b. Anuario estadistico de Pesca 1984. Direccion general de informatica y estadistica, Secretaria de Pesca, Mexico. 338 pp.

Secretaria de Pesca. 1986. Anuario estadistico de Pesca 1985. Direccion general de informatica y estadistica, Secretaria de Pesca, Mexico. 336 pp.

Secretaria de Pesca. 1990. Anuario estadistico de Pesca 1988. Direccion general de informatica y estadistica, Secretaria de Pesca, Mexico. 350 pp.

Secretaria de Pesca. 1991. Anuario estadistico de Pesca 1989. Direccion general de informatica y estadistica, Secretaria de Pesca, Mexico. 125 pp.

Shantha, G., A. Rajaguru, and R. Natarajan. 1988. Incidental catches of dolphins (Delphinidae: Cetacea) along Porto Novo, southeast coast of India. CMFRI spec. publ., no. 40, p. 99.

Sharples, P., K. Bailey, P. Williams, and A. Allan. 1990. Report of observer activity on board JAMARC driftnet vessel R.V. *Shinhoyo Maru* fishing for albacore in the South Pacific Ocean. Working Paper No. 3, Third South Pacific Albacore Research Workshop, Noumea, New Caledonia 9-12 October 1990.

Shimada, H. and H. Nakano. 1992. Impact of the Japanese squid driftnet fishery on salmon shark resources in the North Pacific. Compendium of documents submitted to the Scientific Review of North Pacific Highseas Driftnet Fisheries, Sidney, B.C., Canada, June 11-14, 1991, Vol. 1.

Silas, E.G. and P.P. Pillai. 1982 Resources of tunas and related species and their fisheries in the Indian Ocean. Central Marine Fisheries Research Institute Bull. 32. 175 pp.

Silva, H.M. 1993. A density-dependent Leslie matrix-based population model of spiny dogfish, Squalus acanthias, in the NW Atlantic. ICES C.M. 1993/G:54. 17 pp.

Sivasubramaniam K. 1963. On the sharks an other undesirable species caught by tuna longline. Records of Oceanographic Works in Japan 7(1):73-81

Sivasubramaniam K. 1964. Predation of tuna longline catches in the Indian Ocean, by killer whales and sharks. Bull. Fish. Res. Stn. Ceylon. 17(2):221-236

Sivasubramaniam K. 1987. Some observations on the tuna fisheries in the Indian Ocean, particularly in the central equatorial sub-region, pp. 295-298 in: Collective Volume of Working Documents Presented at the Expert Consultation on stock Assessment of tunas in the Indian Ocean held in Colombo, Sri Lanka 4-8 December 1986.

Sluczanowski, P.R.W., T.I. Walker, H. Stankovic, J. Tonkin, N.H. Schenk, J.D. Prince, and R. Pickering. 1993. SharkSim: A computer graphics model of a shark fishery. pp. 45-48 in: "Shark Conservation". Proceedings of an International Workshop on the Conservation of Elasmobranchs held at Taronga Zoo, Sydney, Australia, 24 February 1991. (Edited by Julian Pepperell, John West and Peter Woon). Zoological Parks Board of NSW.

Southeast Asian Fisheries Development Center 1978. Fisheries statistical bulletin for the South China Sea area 1976. Southeast Asian Fisheries Development Center, Thailand.

Southeast Asian Fisheries Development Center 1979. Fisheries statistical bulletin for the South China Sea area 1977. Southeast Asian Fisheries Development Center, Thailand.

Southeast Asian Fisheries Development Center 1980. Fisheries statistical bulletin for the South China Sea area 1978. Southeast Asian Fisheries Development Center, Thailand.

Southeast Asian Fisheries Development Center 1981. Fisheries statistical bulletin for the South China Sea area 1979. Southeast Asian Fisheries Development Center, Thailand.

Southeast Asian Fisheries Development Center 1982. Fisheries statistical bulletin for the South China Sea area 1980. Southeast Asian Fisheries Development Center, Thailand.

Southeast Asian Fisheries Development Center 1983. Fisheries statistical bulletin for the South China Sea area 1981. Southeast Asian Fisheries Development Center, Thailand.

Southeast Asian Fisheries Development Center 1984. Fisheries statistical bulletin for the South China Sea area 1982. Southeast Asian Fisheries Development Center, Thailand.

Southeast Asian Fisheries Development Center 1985. Fisheries statistical bulletin for the South China Sea area 1983. Southeast Asian Fisheries Development Center, Thailand.

Southeast Asian Fisheries Development Center 1986. Fisheries statistical bulletin for the South China Sea area 1984. Southeast Asian Fisheries Development Center, Thailand.

Southeast Asian Fisheries Development Center 1987. Fisheries statistical bulletin for the South China Sea area 1985. Southeast Asian Fisheries Development Center, Thailand.

Southeast Asian Fisheries Development Center 1988. Fisheries statistical bulletin for the South China Sea area 1986. Southeast Asian Fisheries Development Center, Thailand.

Southeast Asian Fisheries Development Center 1989. Fisheries statistical bulletin for the South China Sea area 1987. Southeast Asian Fisheries Development Center, Thailand.

Southeast Asian Fisheries Development Center 1990. Fisheries statistical bulletin for the South China Sea area 1988. Southeast Asian Fisheries Development Center, Thailand.

Southeast Asian Fisheries Development Center 1992. Fisheries statistical bulletin for the South China Sea area 1989. Southeast Asian Fisheries Development Center, Thailand.

Southeast Asian Fisheries Development Center 1993. Fisheries statistical bulletin for the South China Sea area 1990. Southeast Asian Fisheries Development Center, Thailand.

Sparholt, H., and M. Vinther. 1991. The biomass of starry ray (*Raja radiata*) in the North Sea. J. Cons. int. Explor. Mer, 47:295-302

SPC (South Pacific Commission). 1991. Regional Tuna Bulletin, First Quarter 1991. South Pacific Commission. Tuna and Billfish Assessment Programme. 51 pp.

Springer, S. 1951. The effect of fluctuations in the availability of sharks on a shark fishery. Gulf and Caribbean Fisheries Institutes. Proceedings. 1951. pp. 140-145.

Springer, S. 1960. Natural history of the sandbar shark, *Eulamia milberti*. Fish. Bull. 61:1-38

Springer, V.G. and J.P. Gold. 1989. Sharks in question. Smithsonian Institution Press. Washington, D.C. 187 pp.

Stevens, J.D. 1990. The status of Australian shark fisheries. Chondros 2(2):1-4

Stevens, J.D. 1992. Blue and mako shark by-catch in the Japanese longline fishery off south eastern Australia. Austr. J. Mar. Freshwat. Res. 43(1):

Stevenson, D.K. 1982. Una revision de los recursos marinos de la region de la comision de pesca para el Atlantico centro-occidental (Copaco). FAO, Documentos Técnicos de Pesca, No. 211. 143 pp.

Strasburg, D.W. 1958. Distribution, abundance, and habits of pelagic sharks in the central Pacific Ocean. Fish. Bull. U.S. Fish. Wildlife Serv. 58:335-361

Sudarsan, D., M.E. John and A. Joseph. 1988. An assessment of demersal stocks in the southwest coast of India with particular reference to the exploitable resources in outer continental shelf and slope. CMFRI spec. publ.; no. 40, pp. 102-103.

SUDEPE (Superintendencia do Desenvolvimento da Pesca). 1990. Annual reports of fisherries statistics. SUDEPE, Rio Grande. (not seen, cited in Peres and Vooren (1991)).

Suzuki, S. 1988. Study of interaction between longline and purse seine fisheries in yellowfin tuna, *Thunnus albacares* (Bonnaterre). Bull. Far Seas Fish. Res. Lab 25:73-144,

Swaminath, M., T.E. Sivaprakasam, P.S. Joy and P. Praveen. 1985. A study of marine fishery resources off Krishnapatnam. CIFNET; Cochin (India); 1985; 43 pp.

Taniuchi, T. 1990. The role of elasmobranchs in Japanese fisheries. NOAA Tech. Rep. NMFS 90:415-426

Tetard. 1989a. Species review. Rays. Working document presented at the ICES Elasmobranch Study Group Meeting, Dublin Ireland, April 1989. 13 pp. (ms)

Tetard. 1989b. Species review. Sharks. Working document presented at the ICES Elasmobranch Study Group Meeting, Dublin Ireland, April 1989. 14 pp.(ms)

Tomas, A.R.G. 1987. Chave taxonomica para caçoes eviscerados do sudeste do Brasil. III Reuniao do Grupo de Trabalho Sobre Pesca e Pesquisa de Tubaroes e Raias no Brasil, Fortaleza, Brasil. 28 July - 31 July, 1987. (Abstract)

Uozumi, Y. 1993. Catch at size of albacore caught by Japanese longline fishery in the Atlantic Ocean from 1956-1990. ICCAT Coll. Vol. Sci. Paps. 40(2): 343-353.

van der Elst, R.P. 1979. A proliferation of small sharks in the shore-based Natal sport fishery. Env. Biol. Fish. 4(4):349-362

Varghese, G. 1974. Shark resources of the Laccadive waters. Seafood exp. Jour., 6(1):65-68

Vélez, R., D. Medizával J.J. Valdez and N.A. Venegas. 1989. Prospección y Pesca Exploratoria de Recursos Pesqueros en la Zona Económica Exclusiva del Océano Pacífico. I.N.P. C.R.I.P., México. 179 pp.

Vinther, M., and H. Sparholt. 1988. The biomass of skates in the North Sea. ICES CM 1988/G:48. 26 pp.

Vooren, C. M., and R. Betito. 1987. Caçoes e arraias demersals do rio grande do sul como recursos resqueiros: biomassa, distribuiçao por profundidade, e migraçoes. III Reuniao do Grupo de Trabalho Sobre Pesca e Pesquisa de Tubaroes e Raias no Brasil, Fortaleza, Brasil. 28 July - 31 July, 1987. (Abstract)

Vooren, C. M., M. L. G. de Araujo and R. Betito. 1990. Analise da estatistica da pesca de elasmobranquios demersais no porto de Rio Grande, de 1973 a 1986. Ciencia e Cultura 42(12):1106-1114 (in Port. with Eng. abst.)

Walford, L.A. 1935. The sharks and rays of California. Fish Bulletin No. 45. California Division of Fish and Game. 65 pp.

Walker, T.I. 1988. The southern shark fishery. In: Proceedings of the workshop on scientific advice for fisheries management: getting the message across. Ed. M. Williams. Australian Government Publishing Service, Canberra. 78 pp.

Walker, T.I. 1992. Fishery simulation model for sharks applied to the Gummy shark, *Mustelus antarcticus* Günther, from southern Australian waters. Aust. J. Mar. Freshwater Res., 1992, 43: 195-212.

Warfel, H.E. and J.A. Clague. 1950. Shark fishing potentialities of the Philippine seas. Research Report 15. Fish and Wildlife Service. United States Department of the Interior.

Watanabe, Y. 1990. Report on drop-out observations in drift-net fishing. Third South Pacific Albacore Research Workshop. Inf. Pap. No.2. SPC. Noumeo, New Caledonia 9-12 Oct. 1990. 8 pp.

Wetherall, J. and M. Seki. 1992. Assessing impacts of North Pacific high seas driftnet fisheries on Pacific pomfret and sharks: progress and problems. Compendium of documents submitted to the Scientific Review of North Pacific Highseas Driftnet Fisheries, Sidney, B.C., Canada, June 11-14, 1991, Vol. 2.

Witzell, W.N. 1985. The incidental capture of sharks in the Atlantic United States Fishery Conservation Zone by the Japanese tuna longline fleet. NOAA Tech. Rep. NMFS 31:21-22.

Wood, C.C., K.S. Ketchen and R.J. Beamish. 1979. Population dynamics of spiny dogfish (*Squalus acanthias*) in British Columbia waters. J. Fish. Res. Board Can. 36:647-656.

Woodley T.H. and M. Earle. 1991. Observation on the French albacore driftnet fishery of the Northeast Atlantic. Preliminary report prepared for Greenpeace International. 9 pp.

Yamada, U. 1986. Rajidae. in: Fishes of the East China Sea and the Yellow Sea (O. Okamura, ed.) p. 32-36. Seikai Reg. Fish. Res. Lab. Nagasaki (in Japanese) (not seen, cited by Ishihara 1990)

Yatsu, A. 1989. Cruise report of flying squid survey by the *Wakatori Maru* in July/August, 1989. (Document submitted to the Annual Meeting of the International North Pacific Fisheries Commission, Seattle, Washington, 1989 October.) 20 p. Fisheries Agency of Japan, Far Seas Fisheries Research Laboratory, 5-7-1 Orido, Shimizu, Shizuoka, Japan 424.

Yatsu, A., K. Hiramatsu and S. Hayase. 1993. Outline of the Japanese squid driftnet fishery with notes on the by-catch. International North Pacific Fisheries Commission. Bulletin no. 53(I): 5-24.

Yeh, S.Y., and I.H. Tung. 1993. Review of Taiwanese pelagic squid fisheries in the North Pacific. International North Pacific Fisheries Commission. Bulletin no. 53(I): 71-76.

Yoshimura, H. and S. Kawasaki. 1985. Silky shark (*Carcharhinus falciformis*) in the tropical water of Western Pacific. Rep. Jap. Group Elasmobr. Studies. No. 20, 1985.

Zhow, K., and X. Wag. 1990. Brief review of passive fishing gears and incidental catches of small cetaceans in Chinese waters. Document presented to the International Whaling Commission Workshop on mortality of cetaceans in passive fishing nets and traps. La Jolla, California, October 1990. (SC/O90/21), 13 pp.